T0271161

Technology Innovation

Technology Innovation discusses the fundamental aspects of processes and structures of technology innovation. It offers a new perspective concerning fundamentals aspects not directly involved in the complex relations existing between technology and the socio-economic system.

By considering technology and its innovation from a scientific point of view, the book presents a novel definition of technology as a set of physical, chemical and biological phenomena, producing an effect exploitable for human purposes. Expanding on the general model of technology innovation by linking the model of technology, based on a structure of technological operations, with the models of the structures for technology innovation, based on organization of fluxes of knowledge and capitals, the book considers various technological processes and the stages of the innovation process.

- Explains a novel definition of technology as a set of physical, chemical and biological phenomena producing an effect exploitable for human purposes.
- Discusses technology innovation as result of structures organizing fluxes of knowledge and capitals.
- Provides a technology model simulating the functioning of technology with its optimization.
- Presents a technology innovation model explaining the territorial technology innovation process.
- Offers a perspective on the evolution of technology in the frame of an industrial platform network.

The book is intended for academics, graduate students and technology developers who are involved in operations management and research, innovation and technology development.

Technology Innovation
Models, Dynamics, and
Processes

Angelo Bonomi

CRC Press
Taylor & Francis Group
Boca Raton London New York

CRC Press is an imprint of the
Taylor & Francis Group, an **informa** business

First edition published 2023
by CRC Press
6000 Broken Sound Parkway NW, Suite 300, Boca Raton, FL 33487-2742

and by CRC Press
4 Park Square, Milton Park, Abingdon, Oxon, OX14 4RN

CRC Press is an imprint of Taylor & Francis Group, LLC

© 2023 Angelo Bonomi

ISBN: 978-1-032-37064-4 (hbk)
ISBN: 978-1-032-37075-0 (pbk)
ISBN: 978-1-003-33518-4 (ebk)

DOI: 10.1201/9781003335184

Typeset in Times
by codeMantra

Contents

List of figures

List of tables

Preface

This book concerns technology and technology innovation and its origin is based on experiences and considerations I had about technology since my young age. I was born in a small Italian territory, close to the Swiss board and called Verbano. A territory with a long history of industrialization, started in 1808 with the transfer from Switzerland of entrepreneurs, workers and machineries during the Napoleonic regime. In this way Italy had the first mechanical cotton spinning factory. This type of industry grew in this territory, favorized by presence of hydraulic energy from the rivers of the Alps, becoming in the second half of the XIX century a small industrial district for cotton spinning with more than 5,000 workers and 40 factories. It was also characterized by innovative activities such as the realization in 1891 of the first Italian out-of-town electric power line of 5 km length, from a 500 kW hydroelectric plant close to the mountains, feeding factories with electric instead of hydraulic energy. My first contact with the industrial technology of this territory occurred when I was a young boy 13 years old visiting, accompanied by my father, the local factory producing rayon thread. This artificial silk material was made by treating cellulose foils with acetic acid. The obtained acetalized cellulose was separated and dissolved in acetone, and after this solvent evaporated in a spinneret producing rayon threads. The exhausted mixture was treated with ether to extract the excess of acetic acid, after recovered from the solvent in a distillation column. There were for me two impressive memories of this visit, the first one of fear looking to the high distillation column with its noise of pumps and flows of liquids, the second one of worry looking to a worker substituting filters in a cabin in a saturated atmosphere of acetone. However, these negative emotions did not hinder me the choice, for my secondary studies, to become a technician in chemistry thinking that in fact technologies were at the same time dangerous and interesting. I started in this way my studies in an historical local school, founded in 1886 by the willing of a local industrialist, Lorenzo Cobianchi, died in 1881. His idea was that local industry would need a technical education also for young teenagers, 15 or 16 years old, teaching mechanics, electricity and chemistry already at this young age. In the first year of my course, I was surprised as I had to frequent a workshop, instead of a chemical laboratory, with the task to transform a piece of steel in a bolt with perfect dimensions and planarity using simply file tools and a caliper. I understood much later that it was in fact a good way to teach what is really a knowhow, a knowledge that cannot be learned reading books or hearing lessons in a classroom, but having in fact a great importance in technological activities. After these studies I frequented the University of Milan obtaining a degree of doctor in industrial chemistry. This degree was characterized, in the conception of this university, by the same teaching for the degree in chemistry, but with added courses concerning industrial chemistry and chemical engineering. That in order to make an education well suitable for R&D activities in the development of chemical technologies. I started then my career as researcher at the Battelle Geneva Research Centre. At the beginning my interests were mostly directed to scientific research and R&D project management. However, in a second phase of my career, I decided to start a consulting activity for the promotion of technology innovation in Italian industrial districts, and finally I had the occasion to become a research associate of IRCrES, a Research Institute on Sustainable Economic Growth of CNR, the Italian National Research Council, studying technology innovation and technology development systems of territories. Looking to literature, mostly on relation between technology and industrial economy, I was unsatisfied how technology was considered in many of these studies in respect to my scientific knowledge and technical experience. Technology and technology innovation in economic studies appeared often not clearly defined as in the case of typical concepts used in hard sciences. Technology and innovation definitions were in fact different following encyclopedia, books or articles leading to uncoherent views about the technology innovation process. Furthermore, after more than 50 years of experience in technology innovation and its management, I asked myself whether my past experience was really useful now in the present world of innovation. In fact, it appeared to me that problems and solutions, found in past experience cumulated in my first decades of activity, were not valid to supply solutions in the new environment with different views about innovation, and availability of the modern computers and communication technologies. However, that raised in my thinking the idea that there are nevertheless some fundamentals of technology innovation that will be the same, existing either in the past or in the present situation and even in the future evolution of technology innovation. I thought that it would be of interest to look for these fundamental aspects. The idea was that it is necessary not just only to study how new technologies are developed but rather to think about technology innovation as anybody has never thought until now. From these considerations, I asked myself what it is conserved in the technology innovation activities of homo erectus, half million years ago, in improving the working of stones or using fire for his purposes, and in technology innovation made in the present R&D laboratories by researchers. A possible answer I have found was in taking account the physical nature of technology, as formed by a set of physical, chemical and biological phenomena producing an effect exploitable for human purposes. These aspects are present either in technologies of homo erectus or in the modern technologies. Trying to study technology in this way, it is necessary to take account of an enormous number of physical phenomena that are present in the activity of a modern technology, an extremely difficult task. A possibility to simplify the study of technology with this approach appeared in studies made around the beginning of the 1990s at the Santa Fe Institute in the frame of the science of complexity. In these studies, technology was considered formed by a set of technological operations that are in fact in relation with the physical, chemical and biological phenomena present in a technology. Then technology may be seen as a structure of technological operations, coherent with its physical nature, and its innovation considered as a change of this structure. That raised the question about how this change occurs, and the idea was that this change takes place in structures organizing fluxes of knowledge and capitals, generating in this way new technologies and new knowledge. All that made possible the development of a mathematical model of technology based on its operations, and models of technology innovation occurring in structures organizing fluxes of knowledge and capitals, describing in this way fundamental aspects of technology and its innovation, independently from economic, industrial or social factors. These studies have constituted the object of my previous book on technology dynamics, and the present book on models, dynamics and processes of technology innovation.

Acknowledgments

Many people have contributed to the writing of this book. First of all, I cite Sergio Pizzini, former professor of physical chemistry at the University of Milan, for the discussions had on the nature of technology, its scientific definition and impacts of technology on the environment. Georges Haour, my colleague at Battelle Geneva Research Centre, and after professor of Technology Management at IMD, a business school based at Lausanne (Switzerland), for discussions on evolution of the technology innovation system, and about the formation of an industrial platform network. Mario Marchisio, associate professor in synthetic biology at the University of Tientsin, People's Republic of China, for the discussion on future of synthetic biology. Gabriele Ricchiardi, Director of the Center of Nanostructured Interfaces and Surfaces of the University of Turin, for the discussions on the future of nanotechnologies. Paolo Marenco, President of the association La Storia nel Futuro, for the discussions about the new social types of technological innovations meeting exigencies of people. Finally, I thank my son Pietro for the help in drawing the figures of the book.

Author Biography

Angelo Bonomi obtained a doctorate in industrial chemistry from the University of Milan in 1969. In 1970, he joined the Geneva Research Centre of Battelle Memorial Institute for R&D projects and scientific research in electrochemistry. In 1980, he shifted to studies on technology assessments and trends. In 1988, he joined a startup in France for contract research and generation of startups. In 1993, he was consultant for technological innovation in Italian industrial districts and for urban waste treatments and environmental protection. His major contribution was in the foundation of Consortium Ruvaris, a consortium for cooperation in R&D of about 20 firms producing taps and valves. From 2001 to 2004, he taught technology innovation management at SUPSI (Switzerland). Since 2013, he has collaborated as Senior Research Associate at Research Institute on Sustainable Economic Growth (IRCrES) of the National Research Council of Italy, for studies on technology innovation and territorial innovation systems. He is author of the book *Technology Dynamics* (CRC Press, 2020) and three chapters about models of technology and its innovation in the book, *Innovation Economics, Engineering and Management Handbook*, edited by Prof. Uzunidis and coworkers of the French University of Littoral (ISTE-Wiley, 2021). His research activity is reported on his personal site www.complexitec.org.

List of Abbreviations

AFM	Atomic Force Microscope
AI	Artificial Intelligence
CAS	Complex Adaptive System
DI	Distributed Innovation
EU	European Union
GMO	Genetically Modified Organism
GSO	Genetically Synthetized Organism
ICT	Information and Communication Technologies
IoT	Internet of Things
IPN	Industrial Platform Network
ISE	Innovative System Efficiency
LbyD	Learning by Doing
MRI	Magnetic Resonance Imaging
PET	Positron Emission Tomography
R&D	Research and Development
ROI	Return of Investment
SEM	Scanning Electron Microscope
STM	Scanning Tunneling Microscope
SVC	Startup-Venture Capital
TEM	Transmission Electron Microscope
UNO	United Nations Organization
VC	Venture Capital

Introduction

1

This book describes models, dynamics and processes of technology innovation concerning the technology dynamics [1]. Actually, the importance and novelty of this book is not just about explanation on how new technologies are developed but rather to think about technology innovation in a new way, aiming to reveal its fundamental nature. That is possible by considering technology as a physical phenomenon producing an effect exploitable for human purposes, and new technologies as a combination of preexistent technologies [2]. This view of technology and its innovation concerns fundamentals aspects of technology that are not involved directly in the field of the complex relations existing between technology and the socio-economic system. Actually, technology dynamics has some analogies with Newtonian dynamics that explains the action of force on a body, but neither the practical origin of the force nor which is the purpose of its use. In a similar way, technology dynamics explains the nature of technology, but it does not take account of the purposes for its development, neither it studies the criteria for financing or not financing a development, nor the effects that a new technology has in the socio-economic system, but it studies the fundamental structures and processes that are present in technological activities satisfying human purposes. Nevertheless, despite these limitations, the novelty and interest of this new approach to technology and its innovation concern the explaining of many aspects that cannot be observed in the study of the complex relations between technology and the socio-economic system. That may be useful for a full understanding of these relations. In fact, technology dynamics covers a gap between the studies concerning the development of specific technologies, intended as use of scientific results for this purpose, and the studies on technology limited to the effects that it has on the socio-economic system, or how economy influences the technological change. That leads in fact to the possibility to study technology as an autonomous discipline independent from economic or social factors.

The content of the book may be divided int four parts. In the first part, we discuss the nature of technology and its scientific definition. It follows a description of a model of technology and of its innovation resulting by

DOI: 10.1201/9781003335184-1

activity of structures organizing fluxes of knowledge and capitals. In the second part, we discuss the relation of technology innovation in the frame of various other types of innovations, the importance of the science of complexity in the description of concept, processes and models of technology and its innovation, and finally consider the relations existing between technology and the environment. In the third part, we discuss the application of the model of technology in describing the functioning of a technology and its optimization, and in the description of a model explaining the functioning of a technology innovation system of a territory. In the fourth concluding part, we present some perspectives of evolution of the technological innovation system, the future of new technological sectors with important potential of development, and finally describe some intrinsic dangers of technology evolution.

REFERENCES

1. Bonomi A. 2020, *Technology Dynamics: The Generation of Innovative Ideas and Their Transformation into New Technologies*, CRC Press, Taylor & Francis Editorial Group, London
2. Arthur B. 2009, *The Nature of Technology*, Free Press, New York

The Nature and Definition of Technology

2

2.1 BRIEF HISTORY OF TECHNOLOGY AND ITS INNOVATION

Technology has found its origin in an ancestral activity of homo species preceding the appearance of *Homo sapiens*, and dated back about one or two million years ago, assuming the form of worked stones usable for various purposes. At about half million years ago, it appeared in the homo species the use of fire, and technology innovation activity was clearly shown in *Homo sapiens* since about 75,000 years ago with the improvement of their artifacts. Much later the combined use of fire and knowledge about stones contributed to the development of ceramics and metals. It shall be noted that animals also have a certain use of technology. A chimpanzee is able to break a walnut with a stone and choose the good dimension of the stone. However, the difference is in the fact that homo species have been able to improve their technologies, thereby obtaining more suitable tools for their purposes. Practically, they did not use simply a technology as animals but made technology innovations. It is possible to advance the hypothesis that the presence of a technological activity had formed in a certain way a biological and technological coevolution in the homo species toward the formation of the *Homo sapiens*. A hypothetical example of coevolution and competition between biological and technological evolution might be considered in the case of homo species facing an initial glacial era. In this case, biological evolution might have favored the growth of body hair for cold protection through a Darwinian selection of the population, and a survival of only a minor part of the human species. On the contrary, the availability

DOI: 10.1201/9781003335184-2

of technologies to kill a prey and recover its fur for cold protection had made available a more rapid and efficient solution. This hypothetical example of competition in the evolution shows that technology evolution might be much more rapid and efficient than biological evolution in the survival of the species, but at the same time more dangerous if we consider the present perils of atomic weapons and environmental problems. Actually, although the nature is essential for the existence of living species, it might be considered that the absence of technology innovation in the homo species, probably the appearance of *Homo sapiens*, would not be possible, and homo species would be simply another type primate, characterized by erect walking, that would be existing today or even disappeared as many other types of primate. This coevolution of technology and biology has been also characterized by an important increase of the brain dimension with a probable synergic action between the brain growth and technology development. The characteristic of technology as survival activity also exists of course in the present times, although not often considered in its importance. On the contrary, technology is considered sometimes a danger for the survival of humanity, neglecting the neutral characteristic of technology, and the fact that dangers are actually in the human decisions to use improperly the technology, not in the technology itself. In fact, for example, bows and arrows may be used to kill a prey to ensure survival or to kill a man during a fight. Furthermore, it should be noted that primitive technological innovations were carried out in conditions of absence of writing and perhaps of existence of only primitive languages. That has formed probably a direct knowledge of the use of a technology by the human brain determining what nowadays is called *knowhow*. The consequence is that a technology transfer cannot be assured only in a written or spoken form but needs also imitation, direct experience and learning by doing (LbyD). That will be discussed further in the chapter about scientific models of technology and its innovation.

2.2 PHILOSOPHIC THOUGHT ABOUT TECHNOLOGY

On the search of a suitable definition of technology, it is useful to consider what the philosophers have elaborated about the nature of technology. It is not the aim and possibility of this book to provide in detail philosophical discussions on the various positions of philosophers about technology. However, it may be considered of interest to compare philosophical views about technology with definitions, concepts, processes and structures considered in technology dynamics [1]. In this way, it is possible not only to observe differences or accord with philosophical thought but also to obtain possibly a contribution by philosophic thought to the definition of technology and its innovation. Furthermore, it is interesting to consider that certain philosophers have also written about the dangers of technology, an argument that will be also treated in further chapters considering the relation of technology with the environment and the evolution of technology. For discussions on the philosophical thought about technology, it has been taken as source of knowledge the proceedings of a meeting, organized by the International Federation of Philosophical Societies, held in Brussels on June 20–23, 2002, published with the title "The Philosophers and the Technology" [2]. In these proceedings, it has been noted that the study on the nature of technology has never had a great importance in philosophical thought. It was Aristotle to consider first technology giving it the name, *techne* in the Greek language. Actually, Aristotle gave a meaning to technology as a practical skill, distinguishing this skill of making from the skill of doing [3]. The establishment of such difference might be considered reductive nowadays after the development of modern technologies involving machines with artificial intelligence (AI). Actually, very few philosophers have considered the materiality of technology in a similar view as it is done in technology dynamics. An exception is the case Ernst Kapp, a minor German philosopher who considered technology as a way with which humans increase the capacities of their arms, legs and senses developing machines, means of transport, microscopes, telescopes, etc. [4]. There is also a second American philosopher, Donna Haraway, who has gone further introducing the concept of *cyborg*, the result of a trend of the humanity toward an increased incorporation of technologies in the human body [5]. These two philosophers in fact well express the present technological trends and, in particular, the technological increase of human capacities that, in the second half of the XX century, for the first time in the evolution of technology, has involved not only physical capacities but also intellectual capacities with computers, communication networks and artificial intelligence. Many other philosophers had concepts of technology quite far from those of technology dynamics, sometimes in contrast also with the nature of science. That is the case for example of Hannah Arendt [6] considering modern research, for example in physics, as simple source of data revealed to us as from our instruments of measurement. She gave the example of Eddington's measurements, demonstrating the Einstein's general theory of relativity, and considered these data having the same resemblance in their reality as a phone number linked to its subscriber in a phone list. Arendt was not aware that few decades later the theory of general relativity, confirmed by Eddington's data, was useful to determine a geographical accurate position on earth through geostationary satellites, finding a great application. In fact, the use of this theory makes possible an error of few meters; otherwise, without taking account of the Einstein's theory, the error would be of kilometers. This example shows well that fundamental scientific research and theory, apparently without applications, may become later useful for important technologies of great interest. On the contrary, there is a philosopher, Gilbert Simondon [7], that had a view on technology quite similar to the view existing in technology dynamics. This philosopher discussed the nature of technology in his essay "About the Way of Existence of Technical Objects" [8] affirming: *the invention brings a wave of condensation, of concretization that simplify the (technical) objects charging each structure with a plurality of functions, not only old functions are conserved and better fulfilled, but concretization brings in addition new properties, complementary functions that were not researched, and that might be named "overabundant functions" constituting the class of a true advent of possibilities in addition to the properties expected by the (technical) object.* Considering the term *concretization* not far from the term of *innovation process* used in technology dynamics, and considering the term *function* analogous to *technology* seen as a process, we might consider the *structure with a plurality of function* analogous, in a certain way, to the *structure of new technologies* resulting as a combination of previous technologies [9] adopted in technology dynamics. On the other side, the concept of *overabundant functions, constituting the class of a true advent of possibilities in addition to the properties expected by the (technical) object,* is not far from the potentiality of preexistent technologies in the formation of new technologies also considered in technology dynamics. All that shows that concepts and models of technology dynamics are not special artificial constructs, but may be considered as a natural result of philosophical reflections on technology in action. An important philosophical thought contributing to the understanding of fundamentals of technology and its dangers has been given by the German philosopher Martin Heidegger, and by his scholars Hans Jonas and Hans Georg Gadamer. The work of Martin Heidegger, with his famous essay "The Question Concerning Technology", is considered, from the philosophic point of view, as one of the major contributions to the study of fundamentals of technology. In this essay, the philosopher reported his ideas about the essential nature of technology, but also the intrinsic dangers of technology not necessary linked to its use [10]. Hans Jonas is known for discussions about the dangers of use of technologies in relation with the environment [11] defining the well-known precautional principle [12]. His thought will be presented in detail in the chapter about relation between technology and the environment. Hans-Georg Gadamer had ideas about technology quite different from those of Martin Heidegger, considering in particular the concrete functioning of technology, and existence of dangers of technology because of the modern communication means [13]. The thoughts of Martin Heidegger and Hans-Georg Gadamer about technology and its dangers are reported in more detail as follows.

2.2.1 Martin Heidegger Thought about Technology and Its Dangers

Heidegger position on the nature of technology has been exposed in a series of lectures of this philosopher since 1949, and the essay "The Question Concerning Technology" is contained in his book *Vorträge und Aufsätze* (1954). This essay is difficult to understand especially due to the use of words with meanings that are not employed commonly, and that must be redefined. All that makes difficult in particular the work of translation from German of the used terms, but also difficult by the presence of expressions constituting a thrust, apparently being beyond a comprehension, observations reported by the translator in English of this essay [14]. It is not the task and possibility of this book to enter into a discussion about the complex philosophical thought of Heidegger about technology, and we limit us to find, as much as possible, similitudes and differences of his thought with how technology dynamics considers the nature of technology, and which are his views on the intrinsic dangers of technology. In fact, Heidegger, like technology dynamics, considers technology an activity with a purpose. In fact, *technology is a means to an end* and *technology is a human activity* concluding that *the current conception of technology, according to which it is a means and a human activity, can therefore be called the instrumental and anthropological definition of technology*. However, following Heidegger, *the essence of technology is by no means anything technological*, but a means to see the nature, revealing itself as a potential resource for humans [10]. Differently of a common thought, modern technology is not considered by Heidegger an application of natural science. Technology makes use of scientific knowledge, but this knowledge depends also on technical instruments and their construction, i.e. science would not exist without the existence of technologies that precedes scientific discoveries. This original conception considers technology, and not science, as a fundamental in the relation of humans with the nature, independently of its good or bad developments [10]. Nevertheless, from another point of view, science cannot be considered exclusively depending on availability of technologies. In fact, science makes use of philosophical thought about the nature such as, for example, the ancient Greek thought about the atomistic nature of matter, becoming the base of sciences such as physics and chemistry. Actually, such argument might be however overturned considering that the atomistic view of matter by ancient Greek philosophers was in fact far from the reality. Atom is not an indivisible part of matter, and the knowledge about atomic components is the result of availability of suitable technologies showing the existence of these components. In fact, a scientific description of nature is considered valid when it is confirmed by results of experiments or by observation of the nature with suitable technical instruments, considering in this way sustainable the idea that all science takes origin by existence of technology. From this point of view, following the definition of technology as an activity pursuing a human purpose [9], we might also define science as a technology satisfying the human purpose to know the nature. Concluding it may be observed that if it is true that technology is essential to scientific discoveries, new phenomena discovered by science are essential in the development of new technologies. In fact, science and technology are involved in an intertwining process in which for example R&D is important for both the activities [1]. Following further aspects of Heidegger's thought, the essence of technology is seen as *an eliciting action asking to the nature to liberate a growing force.* An example is that of uranium that produces an atomic energy exploitable for atomic bombs or for nuclear energy production. Actually, the difference of modern technology from classic technology consists, following Heidegger, in the fact that the *man can operate an action that cumulates natural energy in order to have a reinforced use,* resulting, for example, in dams cumulating mountains or rivers water to produce hydroelectric energy, or the cited case of atomic energy exploitation by natural energy contained in uranium. It shall be noted, however, that this Heidegger's view of modern technology is in fact limited by considering only technologies of energetic type, but recent technology developments include also biotechnologies, computer technologies and artificial intelligence in which it is not cumulated energy but it is cumulated knowledge. Heidegger considers the existence of a possible origin of danger of technology beyond its materiality. The danger of technology is *the production itself, seen as a direct activity separated by the man that produces, while the freedom of the man comes by including actions not directly concerning the production, and that may be moral, positive or negative* [10]. In fact, that corresponds, with a simplistic interpretation of Heidegger's words, to the development of technologies with the objective just to develop technology, and in which technology is the major purpose instead of its application. From the thought of Heidegger, concerning technology might have drawn some useful lessons that have interesting correspondences with technology dynamics studies. The first correspondence concerns the Heidegger's definition of technology as an instrumental and anthropological activity pursuing an end. That corresponds to the definition of technology as an activity pursuing a human purpose [9] adopted in technology dynamics. The second correspondence concerns the existence of the role of technology in generating scientific knowledge, affirmed also in technology dynamics discussing for example R&D activity [1]. Finally, the cited alert of Heidegger about intrinsic dangers of technology has implications in technology evolution and will be further discussed in a next chapter concerning this topic.

2.2.2 Hans-Georg Gadamer's Thought about Technology and Its Dangers

Hans-Georg Gadamer was a scholar of Martin Heidegger but with a different thought about technology. Gadamer does not see technology as a massif argument expressing the modern understanding of the human being, but rather considering the concrete functioning of technology. He expressed his ideas on technology in various writings and especially describing its dangers in a writing dated on 1976 [15]. Following the thought of Gadamer, the essence of technology consists in having the means for the production of something making a process safe and controlled. However, he considers that certain characteristics of technology are dangerous for the social organization [13]. In fact, in the past, the human needs were satisfied by the production, but now technology appears creating the needs and relative consumptions. He considers in fact that in the using and entrusting technology there is also a loss of the freedom of action. Furthermore, the technical transformation of the society leads also to a formation of opinions by technological means. The modern technology of information generates a situation that make information an unavoidable need, and the overabundance of its offer compels us to make choices. It is then unavoidable that a society based on technological means of communication leads to a possible manipulation of the public opinion influencing certain decisions, and the possession of technological means of communication is decisive for this purpose. The result is that the increase of availability of information does not correspond to an increase a social reinforcement. Individuals become dependent on a form of life mediated by the technology, but feeling at the same time to be powerless and leading to a reinforcement of their adaptability and apathy toward public affairs. He anticipated also the potentiality of technology in creating new markets, as shown for example by the ideas of Steve Jobs on personal computers [16]. It is of interest how this philosopher has anticipated problems of various nature raised by the development of internet communication and social networks that in fact appeared clearly many decades later [17], arguments that will be discussed further in this book in the section about the intrinsic dangers of technology evolution.

2.3 THE VARIOUS DEFINITIONS OF TECHNOLOGY AND ITS INNOVATION

There are many questions that may be raised about the definition of technology. In fact, technology may be considered a byproduct of science or a means to satisfy economic, military or scientific purposes. In fact, there is a wide range of definitions of what is technology, but not necessarily coherent and indicative of fundamentals of what technology really is. We give here few examples of definitions of technology that may be found in articles, books and encyclopedias:

> *Science of industrial art with the aim to supply to industry theorical and practical knowledge necessary to make and improve the productions.*

> *Whole of machineries, instruments, production processes and applied knowledge of a society in a certain period. It implies the existence of social and productive organizations allowing the realization of these processes. In a stricter scientific way technology is intended as any scientific application to production of good or services.*

> *Whole of norms following the practical development of a human activity.*

It may be noted that the previous definitions are all linked to the relation that technology has with industrial, economic and social activities, but not on fundamental aspects of technology that are existing since the ancestral activity of homo species until the present activities of homo sapiens. In the study of technology dynamics, it has been adopted a more general definition proposed by Brian Arthur in his book on the nature of technology [9]. This author considers technology simply as *a means to fulfill a human purpose*. It is a valid definition that, however, needs to be more detailed in the search of fundamentals of technology. Furthermore, there is another contribution made by Brian Arthur in the study of the nature of technology that concerns the formation of new technologies, i.e. technology innovation. In fact, he considers the formation of new technologies as *a combination of preexisting technologies able to exploit new phenomena discovered by science* [9]. These definitions of technology and its innovation have been adopted in the study of technology dynamics, although it has been considered also that new technologies may be formed by simple combination of preexisting technologies without exploitation of scientific phenomena that may be possibly present in technologies used in the combinatory process [1].

In fact, technology may be considered based on purposes not limited to survival and improvement of human life, but also to extend scientific knowledge and for social purposes not specifically industrial or economic but rather social, meeting exigencies of people. This last purpose, satisfied by technology, is explained in a next chapter discussing the involvement of technology in various types of innovations. On the search of fundamentals of technology and its innovation, it is necessary to take account that such fundamentals should be valid anytime in technological activity. From these constraints, it appears that the constant characteristic of a technology may just lie in its physical nature.

Then technology, considering a scientific approach about its nature, may be seen as a process constituted by *a set of physical, chemical and biological phenomena* occurring in its activity, and resulting in *an effect able to satisfy a human purpose.* This definition may be called scientific as it is a result of an attitude toward technology analogous to that of science toward the study of a natural phenomenon. This approach is discussed in detail in the next chapter concerning a scientific definition of technology and its innovation.

REFERENCES

1. Bonomi A. 2020, *Technology Dynamics: The Generation of Innovative Ideas and Their Transformation into New Technologies*, CRC Press, Taylor & Francis Editorial Group, London
2. Chabot P. Hottois G. 2003, *Les philosophes et la technique*, Librairie Philosophique J. Vrin, Paris
3. Evans D. 2003, "Aristotle on *techne*" in Chabot P. Hottois G. *Les philosophes et la technique*, Librairie Philosophique J. Vrin, Paris, 37–47
4. Timmermans B. 2003, "L'influence hégélienne sur la Philosophie de la Technique d'Ernst Kapp" in Chabot P. Hottois G. *Les philosophes et la technique*, Librairie Philosophique J. Vrin, Paris, 95–108
5. Botbol-Baum M. 2003, Le cyberféminisme d'Haraway ou "'l'utérus technoscientifique'"" in Chabot P. Hottois G. *Les philosophes et la technique*, Librairie Philosophique J. Vrin, Paris, 253–271
6. Roviello A.M. 2003, "La perte du monde dans la technique moderne selon Hannah Arendt" in Chabot P. Hottois G. *Les philosophes et la technique*, Librairie Philosophique J. Vrin, Paris, 203–215
7. Chabot P. 2003, "La philosophie des techniques de Simondon" in Chabot P. Hottois G. *Les philosophes et la technique*, Librairie Philosophique J. Vrin, Paris, 231–242
8. Simondon G. 1959, *Du mode d'existence des objets techniques*, Aubier, Paris
9. Arthur B. 2009, *The Nature of Technology*, Free Press, New York
10. Kemp P. 2003, "La Question de la Technique selon Heidegger" in Chabot P. Hottois G. *Les philosophes et la technique*, Librairie Philosophique J. Vrin, Paris, 163–173
11. Pinsart M.G. 2003, "Hans Jonas: une réflexion sur la civilisation technologique" in Chabot P. Hottois G. *Les philosophes et la technique*, Librairie Philosophique J. Vrin, Paris, 187–202
12. Jonas H. 1979, *Das Prinzip Verantwortung Versuch einer Ethik für die technologische Zivilization*, Suhrkamp, Frankfurt/Main
13. Van Den Bossche M. 2003, "Hans Georg Gadamer et la pensée technoscientifique" in Chabot P. Hottois G. *Les philosophes et la technique*, Librairie Philosophique J. Vrin, Paris, 175–185

14. Heidegger M. 2013, *The Question Concerning Technology* (English translation by W. Lovitt), Harper Perennial Modern Thought, New York

15. Gadamer H.G. 1976, "Hermeneutiek als praktische Philosophie" in *Vernunft im Zeitalter der Wissenschaft*, Suhrkamp, Frankfurt/Main

16. Isaacson, W. 2011, *Steve Jobs,* New York: Simon & Schuster Paperbacks, New York

17. Schultz T. 2015, Tomorrowland: How Silicon Valley Shapes Our Future, *Spiegel Online International*, 03/04/2015

A Scientific Definition of Technology

3

In the previous chapter, we have discussed some fundamental aspects of technology and which are the limits of many definitions of technology found in articles, books and encyclopedias, mostly derived from the relation of technology with its industrial and economic use and social environments. However, considering fundamental aspects of technology existing throughout the evolution of the technological activity, it has been proposed a scientific approach considering the physical nature of technology. It is then possible to define technology as *a set of physical, chemical and biological phenomena producing an effect exploitable for human purposes*. These scientific aspects of technology are independent of industrial, economic and social factors, and this definition cannot be used for the study of relations between technology and the socio-economic system. However, studies derived from this definition of technology, and concerning technology dynamics [1], show that this approach may reveal hidden fundamentals of technology that cannot be seen through the study of the complex relations that technology has with the socio-economic system, but that may have an influence on this system. These fundamentals of technology may be also useful for the explanation of various hidden or poorly considered technological processes. The studies derived from this scientific vision of technology may in fact cover a gap between the studies concerning the development of specific technologies, intended as use of scientific results for this purpose, and the studies on technology limited to the effects that it has in the socio-economic system, or how economy influences the technological change. That leads in fact to the possibility to study technology and its innovation as an autonomous discipline, independent from economic or social factors, explaining fundamental aspects of the dynamics of technology. On search of fundamentals of technology, it is in fact necessary to look to its intrinsic aspects and not to its relations with scientific, economic or social environments. In order to explain better the advantages of a scientific approach to the study of technology, we shall consider the importance of technologies developed without economic purposes, the limits of study of

DOI: 10.1201/9781003335184-3

technology from only an economic point of view and how it is possible to study technology from a scientific point of view.

3.1 IMPORTANCE OF TECHNOLOGIES WITHOUT ECONOMIC PURPOSES

The history of development of technologies shows many examples of this activity dedicated to purposes other than economic, for example for scientific purposes. That is the case of improvement of telescope by Galileo Galilei for astronomic investigations, and modern science is involved in the development of new technologies for large research equipment such as accelerators of particles in the study of matter, or big interferometers for the detection of gravitational waves. However, the most important developments, not linked to priorities of economic nature, are those for military applications that have been of great importance in the XX century for their implication in the development of derived technologies of great economic importance. That was the case in the synthesis of ammonia developed in Germany for the production of explosives during the First World War, because of the unavailability of Chilean guano containing the necessary nitrates. Actually, nitrates are also natural fertilizers and, after the war, there was a great development of this industrial production destined to agriculture. Another important example of technology was developed for military purposes. Finding after important economic applications was the making, during the Second World War, of the atomic bomb with the Manhattan Project [2]. Such project in fact developed a great number of technologies used after for civil applications, the most important being the nuclear production of energy. Probably the most important case of technologies developed for military purposes, finding after an enormous development with great economic and social implications, was the miniaturization of electronic circuits developed during the Cold and Korea Wars for military devices. In fact, this development was not of interest of industry at that time, considering these technologies having high costs and negligible markets [3]. It was Steve Jobs the first to understand the enormous potential of personal computers made by using miniaturized circuits, easy to use by normal people, and not only by professionals, opening enormous markets and further technology developments [4]. Furthermore, it shall be considered that it is also well known that modern networking technologies, backbone of internet, emerged in early 1970 from the US Department of Defense agencies ARPA and DARP [5]. We may consider that, without these early military technology developments, probably most of the present information and communication technologies

(ICT) would not be available or much less developed. All these examples show well how the technological change cannot be exclusively ascribed to economic activities, and that technology activity has its own characteristics not necessarily only linked to economic factors.

3.2 LIMITS OF STUDIES OF TECHNOLOGY FROM ONLY AN ECONOMIC POINT OF VIEW

The study of technology from an economic point of view is covered by a huge literature including also reviews of the various theories developed in order to explain how economy influences the technological change. For this purpose, we have chosen, for a discussion about the limits of economy in studying technology, a synthetic review describing these theories and their capacity to account for technological and institutional co-evolution and change [6]. The choice of this review is justified because, discussing the necessary further research on technological change, it is proposed a complex approach and simulation converging, in a certain measure, to the same used in the study of technology dynamics [1]. In fact, in this review, it is considered that most of the actual progress done on technological change concerns the findings of the empirical studies within and across industries and countries. The main issue ahead should now be the filling of the gap between such empirical increased knowledge and the theoretical constructions. In this respect, the understanding of the whole process, and in particular of the sources of technological change, is hindered by the received theoretical framework. Indeed, theoretical constructions should, at least, not to be in open conflict with the empirical regularities. In fact, it may be observed that empirical knowledge has created a gap with theoretical constructions. About the work that should be done to clarify the source of the process of technological change, the review's author suggests that the complex interaction between economic social and institutional elements, that characterizes such a process, could be modeled and rigorously understood using a complex approach and with the help of simulation, for example by the application of a methodology based on modeling technology [7] and searching in technological landscapes [8]. These studies, cited in the review, are in fact at the base of the descriptions of technology and the technology innovation processes in technology dynamics [1]. Concluding, the limits of the study of technological change, from an economic point of view, may be ascribed also to the fact that technological change or technology innovation in economic studies is based on a concept of technology not clearly defined, carrying out a whole range of meanings not necessarily coherent. In fact, in accord with Brian Arthur's thought about the nature of technology [9], there is no agreement on what the word "technology" means, no overall theory on how technology comes into being, no deep understanding of what "innovation" consists of, and no theory of evolution of technology. In technology dynamics, there is the advantage consisting in a clear definition of technology, based on a scientific point of view, that makes possible also the separation of the physical nature of technology from the purposes of its use. That allows the development of a rigorous coherent general model of technology, using concepts of the science of complexity, that may have a mathematical description, and that are useful to explain technological processes and the activity of the organizational structures for technological innovation [1].

3.3 THE STUDY OF TECHNOLOGY FROM A SCIENTIFIC POINT OF VIEW

The study of technology as a set of physical, chemical and biological phenomena, in terms of relations of these phenomena within and across their sets, is extremely complex for the enormous number of phenomena that normally shall be considered. The science of complexity may simplify the study considering technology as a set of *technological operations* that in fact are related to the physical phenomena of a technology. That allows the simplification of the study of technology in terms of relations within an operation or among operations, and not among the enormous number of phenomena. Furthermore, considering technology as a process, it exists in a technology a sequence of these operations. Such sequence may be represented in the form of a graph structure in which technology operations are represented as arcs oriented with time. Such structure is similar to the graph representation of tasks used in project management. In order to well explain the relation between the set of physical phenomena of a technology and the definition of technological operation, we may describe as example a simple technology such as that of heat treatment.

This technology, used for example to harden a metallic tool, starts with heating and involves some physical or chemical phenomena able to supply heat. The metallic tool increases its temperature until reaching a value that causes the starting of a phenomenon of change of its crystalline structure. At this point, it is necessary to maintain the reached temperature for the necessary time to obtain the complete transformation of its crystalline structure, and at the same time to maintain the temperature through a physical or a chemical phenomenon able to supply heat. When the entire change of the crystalline structure of the tool is obtained, it is necessary to cool rapidly the tool in order to avoid the transformation of the obtained crystalline structure into the previous structure. That is obtained by use of a physical phenomenon able to subtract rapidly heat from the tool. The obtained tool, with a new crystalline structure, harder than the original structure, will be more efficient for cutting, and that is the exploitable effect of the set of physical phenomena characterizing the technology of heat treatment.

In this example, we note that the technology of heat treatment may be considered consisting of three technological operations and concerning: heating, maintaining temperature for a certain time and cooling, each including a set of various phenomena concerning heating, cooling and change of the crystalline structure. The heat treatment technology may be considered describable by a graph structure consisting of a sequence of three time-oriented arcs representing, respectively, the heating operation, the maintenance operation and the cooling operation. The graph representation of this technology in terms of its operations consisting of sets of physical and chemical phenomena is reported in Figure 3.1. Furthermore, such representation of technology as a graph structure of technological operations may suggest that *technology innovation may be represented by a change of this structure.* That is by additions, substitutions, eliminations and changes of position of technological operations of a preexistent technology in order to form a new technology. Concluding, technology, seen from a scientific point of view, leads to a structure in terms of operations constituting in fact the bricks useful for the development of a general model

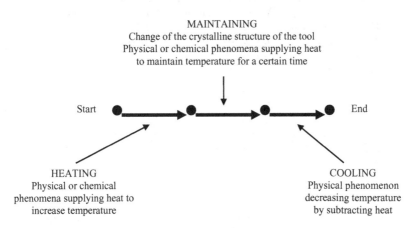

FIGURE 3.1   Graph representation of the heat treatment technology with its physical and chemical phenomena.

of technology. A scientific definition of technology, with the description of its structure, leads, using suitable mathematical tools, to the development of a general model of technology simulating its functioning, and enabling the optimization of its operative conditions. Furthermore, it may explain some important technological processes existing in its activity, including the transfer of technology and explanation of what is the knowhow.

REFERENCES

1. Bonomi A. 2020, *Technology Dynamics: The Generation of Innovative Ideas and Their Transformation into New Technologies*, CRC Press, Taylor & Francis Editorial Group, London
2. Rhodes R. 1986, *The Making of the Atomic Bomb*, Simon & Schuster, New York
3. Giarini O. Loubergé H. 1978, *The Diminishing Returns of Technology: An Essay on the Crisis in Economic Growth*, Pergamon Press, Oxford
4. Isaacson W. 2011, *Steve Jobs*, Simons & Schuster, New York
5. Feldman P. Francis J. 2002, "The Entrepreneurial Spark: Individuals Agent and the Formation of Innovative Clusters" in Curzio A. Fortis M. *Complexity and Industrial Clusters*, Physica Verlag, Heidelberg, 195–212
6. Di Maio M. 2003. *Explaining Technological Change: A Comparative Survey*, Department of Applied Economics, (PhD thesis work), Università degli Studi di Siena, Siena, Italy
7. Auerswald P. Kauffman S. Lobo J. Shell K. 2000, The Production Recipe Approach to Modeling Technology Innovation: An Application to Learning by Doing, *Journal of Economic Dynamics and Control*, 24, 389–450
8. Lobo J. Macready G.W. 1999, Landscapes: A Natural Extension of Search Theory, *Santa Fe Institute Working Paper*, 99-05-037
9. Arthur B. 2009, *The Nature of Technology*, Free Press, New York

Scientific Models of Technology and Its Innovation

4

4.1 ELABORATION OF A SCIENTIFIC MODEL OF TECHNOLOGY INNOVATION

In the elaboration of scientific models of technology and its innovation, it is necessary to consider at the same time technology and technology innovation being in fact like two faces of the same coin. In the previous chapters on the nature of technology and its scientific definition, it has been established a certain number of concepts that are at the base of a scientific model of technology innovation. These concepts are based on a vision of technology as a set of phenomena of physical, chemical and biological nature producing an effect exploitable for the fulfilling of human purposes. Furthermore, new technologies may be seen as a combination of preexistent technologies [1], possibly exploiting new phenomena discovered by science [2]. Concluding, we have established that technology may be described as a set of technological operations [3], assuming a time-oriented structure representable as a graph, and technology innovation as a change of such structure [1].

Considering technology having a graph structure consisting of technological operations, and technology innovation as a change of such structure, it is then possible to develop a general model of technology with a mathematical

DOI: 10.1201/9781003335184-4

description valid for a single technology [3], and for a set of technologies pursuing the same purpose [1]. However, the definition of technology innovation, as a change of the structure of a technology, needs to be explained further in order to describe how this change occurs. For this purpose, in technology dynamics, technology innovation has been considered an activity occurring in structures organizing fluxes of knowledge and capitals with the objective to develop new technologies. The historical origin of the first organization of fluxes of knowledge and capitals for the generation of new technologies may be ascribed to the German dye industry, around 1870, with the establishment of industrial R&D laboratories, realizing a continuous generation of innovations, and not simply separated developments of single ideas from inventors [4]. This new way to generate new technologies evolved in the second half of the XX century by forming a system consisting of startups developing new technologies, financed by venture capital (VC) that, differently from industrial capital, develops technologies for their sale and not for their exploitation. Furthermore, at the beginning of the XXI century, it appeared a third way to organize fluxes of knowledge and capitals in the frame of what it is called an industrial platform [1]. This last system is based on continuous relations between suppliers and users of technologies in order to increase the availability of knowledge, and then to increase the generation of new technologies. These three types of structures organizing fluxes of knowledge and capitals shall not be considered alternative but inclusive. In fact, the activity of startups includes R&D projects, and startups and R&D projects may be present in the structure of an industrial platform. The models of technology and of the structures organizing fluxes of knowledge and capitals for technological innovations, including their possible mathematical descriptions, have been presented in detail in a previous book on technology dynamics [1]. We summarize here briefly the general model of technology, the main technological processes including the transfer of technology and knowhow, the models of the organizational structures for technology innovation and the stages of technology innovation. Finally, are established the principles and the main characteristics of a general model of technology innovation starting from the general model of technology.

4.2 THE MODEL OF TECHNOLOGY

The mathematical model of technology, based on a description of technology as a set of technological operations, is in fact derived, from the mathematical point of view, by analogies with a previous model on interaction of genes in a biological entity [5]. This mathematical model of technology has been originally presented in 1996 at 71st Annual meeting of the Western Economic Association in San Francisco [6] for explaining the activities of LbyD. This model was first published in 1998 as working paper of the Santa Fe Institute, and then in 2000 in the *Journal of Economic Dynamics and Control* [3]. The model is valid for a single technology but it has been after extended in technology dynamics for the description of sets of technologies having the same purpose, and considering not only the economic efficiency but also other types of efficiency of a technology [7]. In this model, the technological operations are considered to be controlled by a certain number of parameters, and each parameter may assume a certain number of discrete values or choices in a determined range. Taking for example the simple technology of heat treatment used previously to describe the physical and chemical phenomena existing in a technology, it is possible to describe the heat treatment technology as consisting of three operations, heating, maintaining temperature and cooling, representable in a graph with three arcs, oriented with time, corresponding to the technological operations that, through their parameters, control the technology. Such representation is reported in Figure 4.1. It shall be noted that technological operations have themselves the nature of a technology, and then a technology may be described using a fine or gross structure of the operations. For example, in Figure 4.2, it is presented the gross structure of the technology of production of faucets and, for example, in this case, the operation of chromium plating may be detailed as consisting of a sequence of three sub-operations of degreasing, nickeling and chroming. The mathematical development of this model allows for the definition of various concepts about technology such as the technological space, the technological landscape, the space of technologies and finally a concept of technology ecosystem. The detailed mathematical description of this model may be found in Appendix 1 of the book on technology dynamics [1].

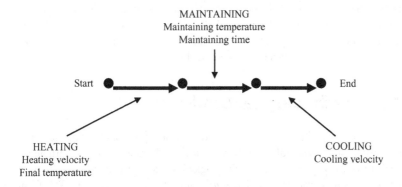

FIGURE 4.1   Structure of the technology of heat treatment.

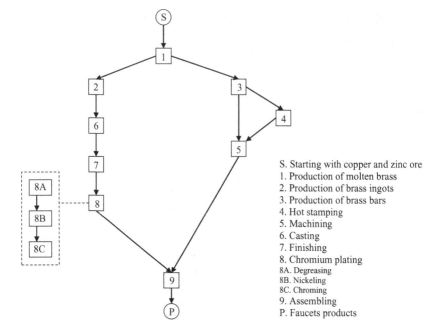

FIGURE 4.2   Structure of technology of production of faucets.

4.2.1 Technological Space

Considering the operations of a technology with their parameters and their respective values or choices existing in a certain range, it is possible to obtain, by a combinatory calculation, all the possible configurations or recipes that characterize a technology. These recipes may be represented by points in a discrete multidimensional space called *technological space* [3]. The Hamming distance between two points in this space represents a measure of the difference between two recipes of the same technology.

4.2.2 Technological Landscape

A technology and its operations may have various types of efficiency (economic, energetic, environmental, etc.). If we associate the scalar value of efficiency to each recipe represented in a technological space, we transform this space in what it is called a *technological landscape* of a technology [3]. The form of the technological landscape depends on the nature of the considered efficiency, following the considered purpose of use of the technology. The search of an optimal recipe for the use of a technology may be seen as an exploration of its technological landscape looking for a "peak" of efficiency. Such landscape has been the object of further studies concerning the optimal conditions of efficiency [8], in terms of search of an optimum by an adaptive explorative walk [9], in a study on recombinant search in the invention process [10] and in technological search in landscapes mapped by scientific knowledge [11].

4.2.3 Space of Technologies

The technological space is useful to describe a single technology; however, when discussing of competition or evolution of various technologies, it may be useful to have a space representing all technologies satisfying the same purpose. The structure of a technology may be represented by a graph that may be described as a matrix. It is then possible to represent all technologies as points corresponding to the different matrices in a discrete multidimensional space called *space of technologies* [7]. The Hamming distance between two points in the space of technologies represents the difference, or the *degree of radicality*, of a new technology in respect to a preexistent technology. Innovations may be considered radical if the Hamming distance between the two technologies is great, or incremental if it is small. In this way, the space of technology defined by the model offers a special view of what it has been defined as natural trajectories of technical progress [12] in the frame of technological paradigms [13].

4.2.4 Technology Ecosystem

The model of technology may be extended in order to describe, in terms of sets, what it may be called a *technology ecosystem*. In fact, from the previous definitions of the model, a technology is considered as a set of technological operations, and the space of technologies as a set of technologies having the same purpose in which each technology is also a set of technological operations. It is then possible to describe a technology ecosystem through a universal set U consisting of all the technological operations of all the represented technologies. In this case, a technology will be consequently a subset of the set U of the technological operations of the ecosystem. Such description of the technology ecosystem may explain various aspects of generation of technologies and their interactions. Considering for example the formation of new technologies as result of a combinatory process of preexisting technologies [1], the consequence is that a new technology is formed by inclusion of all the subsets of operations used in the combinatory process of its formation, being at the same time a subset of *U*. The space of technologies may be also considered a subset of *U* consisting of all sub-subsets of operations of the technologies having the same purpose. It is interesting to note that, while it is highly improbable the existence of two identic spaces of technologies having different purposes, on the contrary it is possible to have two technologies with different purposes with all or partial common technological operations. On the other side, in the technology ecosystem, a technology may influence another technology acting on the externalities influencing the other technology. All that may be seen schematically in Figure 4.3 representing the universal set *U* of all technological operations. In this set for example, the technologies A, B and C are represented as subsets of *U*. The technology A may influence the technology B as indicated by the arrow. On the other side the technologies B and C may have common technological operations represented by the overlapping of their subsets in the universal set *U*. The existence of overlapping of technological operations in two technologies has consequences in the development of new technologies. In fact, the knowledge about previous existent technological operations may make the development of a new technology easier and less expensive. Concerning the interactions among technologies, occurring by influence of the externalities, this fact may generate advantages or disadvantages for a new technology appearing in the ecosystem. For example, in the case of introduction of a new environmental technology in the ecosystem, this technology could have disadvantages in the case that the ecosystem is constituted prevalently by conventional technologies without common technological operations, on the contrary, advantages in the case that the ecosystem is

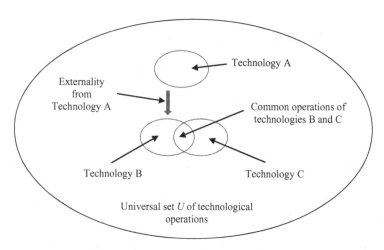

FIGURE 4.3 View of the technology ecosystem with interactions among technologies.

already constituted prevalently by environmental technologies with common technological operations. This interaction will be discussed further in a next chapter on relation between technology and the environment. Concluding, a technology ecosystem expressed as universal set U of operations may present two particular interactions among technologies: the first one in terms of influence of a technology on the externality of another technology, the second one on the possible existence of an overlapping of operations in two technologies that influence the development of a new technology in which some operations are already well-known in the functioning of a preexistent technology.

4.3 TECHNOLOGICAL PROCESSES

The model of technology may explain various types of processes observed in technological activities. These processes may be characteristic of a single technology, or by a set of technologies pursuing the same purpose, and are explained in detail in the study of technology dynamics [1]. We report here only the main technological processes that are directly linked to technology innovations. These processes are consequent by externality or intranality effects on a technology, and also by the process of ramification of an initial radical technology generating new technologies that may be observed in the space of technologies.

4.3.1 Externality Effect

This important effect occurs during the use of a technology, and it is the result of externalities such as changes in raw materials, variation in costs and new regulations to be complied. These externalities may modify the technological landscape reducing possibly the efficiency of the used recipe [3]. It is then necessary to search a new optimal recipe by exploring the new landscape through an activity for example of LbyD, or even to change the structure of the technology realizing an innovation normally of incremental type with a low degree of radicality.

4.3.2 Intranality Effect

The intranality effect consists in the fact that, changing the parameter values of an operation in order to improve its efficiency, that may influence the efficiency of other operations of the technology [3]. Consequently, the optimization of the entire technology shall be obtained by a process of tuning, changing the various parameter values or choices of the operations of a technology. The intranality effect exists also in the change of an operation in the structure of a technology. That may influence the efficiency of other operations of the structure. In this case, the operative conditions of other operations of the structure shall be changed, even introducing possibly further modifications of the structure, necessary for an efficient use of the innovation [1].

4.3.3 Ramification of Technologies

This important technological process occurs when a new technology with an important radical degree appears in the space of technologies, and triggers the formation of other technologies that represent improvements, diversifications or alternatives to the initial radical technology [1]. Technology ramification is characterized normally by a decrease of the radical degree of the formed technologies and by an increase of their number with the development of the ramification. An indirect demonstration of existence of technology ramification may be found by studying the formation of patents from an initial patent of a radical innovation. That is the case for example of the topological evolution of patents from an initial patent covering computerized tomography from 1975 to 2005 [14].

4.4 TECHNOLOGY TRANSFER AND KNOWHOW

With the term *technology transfer* are commonly intended in fact two different processes, the first one concerns the bringing in use a new technology after its development, the second one the transfer of a used technology to another location or in more general way the transfer of technology from an expert to a newcomer. These two processes present similarities but also differences and both need the development of a knowledge named *knowhow* necessary to operate the transferred technology [1]. The knowhow may be explained by the model of technology as follows: a technology is normally influenced by externalities, often with limited effects, that however change the technological landscape. The operator in most cases changes simply the parameters values in order to restore optimal conditions using his knowledge of technological or scientific nature as well accumulated experience. As the effects of externality may be of different types and appear and disappear many times, the necessary changes to maintain optimal conditions of operation for the technology are memorized by the operator and constitute his knowhow of the technology. Such complex knowledge cannot be transferred simply by oral or written instructions but needs imitation and LbyD activity for the operator willing to acquire the knowhow. In fact, a technology cannot be considered a deterministic system for which it is possible to give a complete description of its operative conditions, as it operates in a chaotic environment undergoing to unpredictable externality effects influencing its technological landscape. In the case of transfer of technology by bringing in use a new technology, it is necessary to solve further problems of knowhow associated to the scale up of production from pilot plants or of products from prototypes [1].

4.5 MODELS OF THE ORGANIZATIONAL STRUCTURES FOR TECHNOLOGY INNOVATION

As previously discussed, the model of technology considers technology innovation as a change of the structure of a technology. However, such change shall be described with more details in order to explain fully the process of technology innovation. As previously noted, technology dynamics considers technology innovation as the result of activity of structures organizing fluxes of knowledge and capitals. These organizational structures, as previously cited, are: *the industrial R&D system, the startup-venture capital (SVC) system and the industrial platform system* [1]. These three organizational structures are briefly described as follows.

4.5.1 The Industrial R&D System

The R&D activity may be considered occurring in a structure making innovations by organizing two fluxes respectively of knowledge and capitals [15], and a model of the R&D process is reported schematically in Figure 4.4. It consists in an input of financed projects, partly abandoned and other successful in producing new used technologies. Knowledge generated by both successful and abandoned projects, combined with technical, scientific or other knowledge, generates innovative ideas that may be transformed in R&D project proposals submitted and then selected for financing. New technologies enter in use with availability of industrial capitals and producing returns of investments.

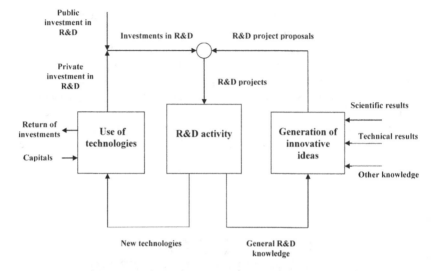

FIGURE 4.4   Schematic view of the industrial R&D system.

Industrial capitals, combined possibly with public funds, make available investments for financing selected R&D projects proposals closing the knowledge and capital cycles. The knowledge generated by the R&D activity by both successful and abandoned projects is for the model the driving force for the generation of innovative ideas for new R&D projects. A mathematical simulation of the R&D model shows that it is necessary a critical minimum number of R&D projects (i.e. investments), in fact a threshold, to obtain statistically at least one new usable technology and even more for a successful technology generating economic growth [16]. Consequently, the economic growth of a territory does not depend actually simply on R&D investments in general, but rather on the efficiency of exploitation of available knowledge in generation of innovative ideas, on adopted strategies about availability of capitals, and a suitable industrial organization [1].

4.5.2 The Startup-Venture Capital System

This SVC system consists of companies called startups financed by VC. Startups differ from R&D projects because they do not make only R&D but also business model developments suitable for the developed technology. Startups have as objective the selling of the developed technology with its business or in collecting capitals to become an industrial company. VC has strategies completely different from those of industrial capital financing R&D projects as it develops technologies for their sale and not for their exploitation, that in an operation called *exit*. The SVC system includes a financial cycle, reported in Figure 4.5, starting with startup projects proposals to VC, selection of financed startups that in part reach an exit, and in part are abandoned. The return of investment (ROI) to VC obtained by selling startups may be used to refinance new startups. A mathematical simulation of the cycle shows that, if ROI is high enough, covering not only the made investments, there is the formation of an autocatalytic development of financing capabilities for new startups boosting technological and economic growth [17]. The necessity of a high ROI, in order to close positively the financial cycle of VC, makes this system particularly suitable for the development of radical innovations, potentially able to give higher returns than in the case of the R&D industrial system. In fact, the experience shows that a successful VC strategy is based on selection of startups with high ROI potential and excellent experienced teams instead of a selection based on the technological probability of success of the startup project, and that needs a suitable knowhow in financing and monitoring startups activities [1]. Finally, it shall be considered that, besides its financial cycle, there is for

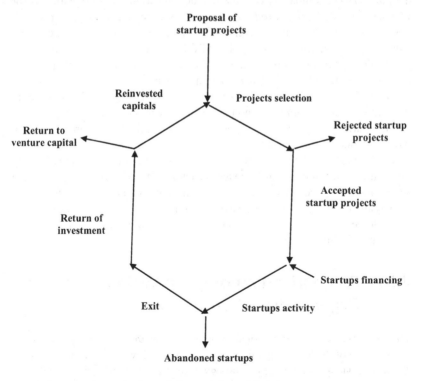

FIGURE 4.5   Schematic view of the SVC financial cycle.

the SVC system also a knowledge cycle, similar to that of R&D, consisting in knowledge formed in either successful or abandoned startups useful for the creation of new startup projects, but consisting not only of technical but also of business model knowledge. It shall be noted that both knowledge and capital of the SVC system may trigger an autocatalytic growth of innovations when certain threshold levels of knowledge and capital availability are overcome.

4.5.3 The Industrial Platform System

The industrial platform system realizes a new form of technology development based on continuous relations between offer and demand of new technologies, and it may include R&D projects and startups activities [1]. In fact, industrial platforms derive from a general platform system largely applied in social and economic activities [18], and its industrial version has known a first diffusion in the field of ICT. The basic actors of an industrial platform are: the *owners* that represent the proprietors assuring the vision of the platform; the *partners* i.e. companies with a continuous relation with the platforms for various services; the *peer producers* that may be companies, startups, research laboratories with discontinuous relations with the platforms supplying complementary services, new technologies and R&D projects; the *peer consumers* i.e. companies or customers interested to buy, in a continuous relation, products or services of the platforms. Such relation of exchange of information may be automatic for example with products in internet of things (IoT). Externally there are the *stakeholders* of the platforms making controls and regulations for the platforms and interested possibly to their growth and prosperity. An industrial platform is characterized by a strong exchange of knowledge, in particular between the developers and the users of a new technology, and by monetary transactions among the various actors of the platform in which there are payments for products or services by peer consumers to the platform, and the payments by the platform to partners and peer producers for services, new technologies or for their development. The structure of the platform is reported schematically in Figure 4.6.

4.5.4 Comparison among the Various Organizational Structures

The three types of organizational structures, described previously, represent an evolution of the technology innovation system following the various ways to exploit knowledge, to finance technology developments in relation with the various degrees of radicality of the innovation. There is in particular a difference

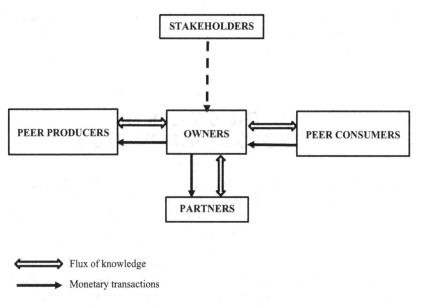

FIGURE 4.6 Schematic view of the industrial platform system.

between the investment strategies of the industrial financing of R&D and the VC financing of startups. In fact, as we have previously noted, while industrial financing is oriented vs. the use of the developed technology, the VC financing is oriented vs. the sale of the developed technology. Furthermore, while the industrial capital decides the amount of financing of new R&D projects following various factors not necessarily linked to the results of financed technologies, the VC exploits the positive results of its investments to finance further startups generating an autocatalytic increase of availability of further capitals for startups. Following the discussed aspects of these two organizational structures, it appears, as previously cited, that the industrial R&D projects system is more suitable for innovations of incremental type while the SVC system is more suitable for radical innovations in which the high potential of ROI would compensate largely the higher probability of investment failure in startups [1]. The industrial platform system is completely different from the two other organizational systems as its strategy is based principally on the increase of knowledge available for innovations, and not on financing. That is obtained in particular by the exchange of knowledge between the platform and the peer consumers, and its efficiency derives from the great availability of knowledge favoring the combinatory nature of innovation, in accord with the previous observations on the development of new technologies. The diffusion of industrial platforms might change technological competition among firms into competition among platforms, while the offer of technology developments by research entities might shift offer toward firms to offer toward platforms becoming peer producers [1].

4.6 THE STAGES OF TECHNOLOGY INNOVATION

Considering the organizational structures with the technological processes described previously, we may give now an articulated view of technology innovation stages taking account of the various possible paths and conditions followed by an innovative idea from its generation to its transformation into a new technology that finally enters in use. From the point of view of technology dynamics, the generation of new technologies may occur through one of the three organizational structures for innovation, i.e. the industrial R&D system, the SVC system and the industrial platform system. The choice of the system depends on various factors described previously in comparing the various organizational structures, and including the degree of radicality of the innovation, the financing strategies and the importance given to knowledge in improving or generating new technologies. The temporal sequence of a technology innovation may be divided in every case in three stages: the generation of innovative ideas, the development of the innovative ideas until the formation of new technologies, and the generation of innovations during the use of the technologies. The development stage consists of a first step concerning feasibility, a second step concerning the determination of performances and economy of the innovation, followed by a last step concerning industrialization. These stages and steps are represented schematically in Figure 4.7 and are described as follows.

4.6.1 Generation of Innovative Idea for the Technological Innovation Process

The generation of an innovative idea is fundamentally a combinatory process involving preexisting technologies with exploitation or not of new or never exploited phenomena discovered by science. This generative process may be the result of individual creativity [19], or may emerge by generative relations among various actors interested in the innovation [20]. The combinatory process includes normally also general scientific and technical knowledge as well

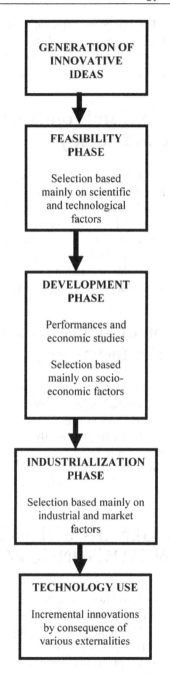

FIGURE 4.7   The stages of technology innovation.

as other types of knowledge. Important exploitable sources of knowledge are coming from the activity of the organizational structures and concerning successful or abandoned projects or startups, and by LbyD activities occurring during the use of a technology.

4.6.2 Development and Formation of a New Technology

This development may occur generally in one of the three different organizational structures constituted by the R&D project system, the SVC system and the industrial platform system, and concerns the verification of the feasibility of the innovative idea, important specially in the case of exploitation of new phenomena discovered by science, followed by the determination of the performances of the technology and estimation of its economy, through studies on prototypes or pilot plants in order to verify the validity of the technology, and finally by the industrialization step making the technology ready for its use [1].

4.6.3 Generation of Innovations during the Use of a Technology

During the use of a technology, there are externalities and intranalities that influence the efficiency of the technology. That leads to the search of new better conditions of operation or even making some changes in its structure and then generating a new technology normally of incremental type. In less frequent, the case of birth of a new idea leading to the development of a more efficient alternative radical technology [1].

4.7 A GENERAL MODEL OF TECHNOLOGY INNOVATION

The previous discussions on technology, technological processes, organizational structures for technology innovation and the stages of development of a technology innovation may lead to an integration of the various models forming a general model of technology innovation.

Staring from the model of technology, we have defined technology innovation as a change of the operational structure of the technology. This change may be detailed considering various models of structures organizing in different ways fluxes of knowledge and capitals in generating new technologies. In this way, it is possible to define a general model of technology innovation, consisting of the models of technology and by the structures for technology innovation. That may be summarized by a certain number of definitions, principles and descriptions as follows:

Definition of technology and its innovation:

1. Technology is an activity fulfilling a human purpose.
2. Technology may be considered consisting of a set of phenomena of physical, chemical and biological nature producing an effect exploitable for the fulfilling of human purposes.
3. New technologies are formed by new combinations of previous technologies exploiting or not exploiting new phenomena discovered by science.

Technology and its innovation described through a model:

1. Technology may be described in form of a model based on sets of physical phenomena characterizing a technology that may be represented by a set of technological operations assuming a time-oriented graph structure.
2. Technology innovation may be seen as a change of a preexistent technology structure.
3. Technological operations have the nature of a technology and then considered also a set of physical, chemical and biological phenomena contributing to the formation of a technology.
4. As technological operations have also a technological nature, a technology may be described with a gross or fine structure depending on the detail with which are represented the operations.
5. Technology may be described through a model based on its configuration defining: a technological space, a technological landscape, a space of technologies having the same purpose, and an ecosystem that may be described as a universal set of all used technological operations.
6. In this technology ecosystem a new technology may appear as a subset in the universal set of operations containing the combination of subsets of operations of preexistent technologies. A technology of the ecosystem may influence the externality of another technology with consequences on its efficiency. Furthermore,

two technologies may have an overlapping consisting of the same operations, that favoring the development of a new technology that has common operations with a previous existing technology.

7. Any technology is characterized by a degree of radicality that measures its difference in respect to a previous technology in the space of technologies with the same purpose, and corresponding to a distance in this space.

8. The technology activity consists of technological processes, explained by the general model of technology.

9. The use of a technology requires a knowledge, called knowhow, that is not transferable completely in a spoken or written form but necessitates imitation and direct experience.

10. The technology innovation process may be considered consisting of three phases: the generation of the innovative idea, the development of the new technology and the use of the technology as source of new technologies.

Technology innovation described through models of structures organizing fluxes of knowledge and capitals:

1. The change of the structure of a technology, and then its innovation, occurs in structures organizing fluxes of knowledge and capitals for the generation of new technologies and new knowledge.

2. These structures are: the industrial R&D projects system, the SVC system and the industrial platform system.

3. The SVC system differs from the R&D project system as venture capital finances development of new technologies for their sale and not for their exploitation.

4. The industrial platforms system differs from the R&D project and SVC systems being based on generation of knowledge through the relations among its actors, and not on different financial strategies.

Following the definitions and the characteristics of technology innovation described previously, it shall be noted, as already discussed in the chapter on the scientific definition of technology, that this general model does not take account of the complex relations existing between technology and economy, in particular it does not take in considerations the amount of available investments, or the financing or not financing a technology innovation. Neither it enters discussion about the economic impact of new technologies nor how new technologies influence economy, but it is limited to consider that new technologies may have different economic impacts, and that a high degree of radicality of a new technology entering in use may be normally associated to a probable high economic return. Nevertheless, the knowledge of the fundamental aspects of technology, which are independent of economic factors, may be useful in technology management and establishment of policies for technological innovation. This knowledge may be also useful for the development of new methods and strategies necessary to conserve the efficiency of innovation activities faced to future expected radical changes of technological activities.

REFERENCES

1. Bonomi A. 2020, *Technology Dynamics: The Generation of Innovative Ideas and Their Transformation into New Technologies*, CRC Press, Taylor & Francis Editorial Group, London
2. Arthur B. 2009, *The Nature of Technology*, Free Press, New York
3. Auerswald P. Kauffman S. Lobo J. Shell K. 2000, The Production Recipe Approach to Modeling Technology Innovation: An Application to Learning by Doing, *Journal of Economic Dynamics and Control*, 24, 389–450
4. Basalla G. 1988, *The Evolution of Technology*, Cambridge University Press, Cambridge
5. Kauffman S. 1993, *The Origin of Order. Self-Organization and Selection in Evolution*, Oxford University Press, New York
6. Auerswald P. Lobo J. 1996, *Learning by Doing, Technological Regimes and Industry Evolution,* 71st Annual Meeting of the Western Economic Association, San Francisco, CA
7. Bonomi A. Marchisio M. 2016, Technology Modelling and Technology Innovation, How a Technology Model May Be Useful In Studying the Innovation Process, *IRCrES Working Paper* 3/2016
8. Kauffman S. Lobo J. Macready G.W. 2000, Optimal Search on a Technology Landscape, *Journal of Economic Behaviour and Organization*, 43, 141–166
9. Lobo J., Macready G.W. 1999, Landscapes: A Natural Extension of Search Theory, *Santa Fe Institute Working Paper*, 99-05-037
10. Fleming L. Sorenson O. 2001, Technology as a Complex Adaptive System: Evidence from Patent Data, *Research Policy*, 30, 1019–1039
11. Fleming L. Sorenson O. 2004, Science as a Map in Technological Search, *Strategic Management Journal*, 25, 909–928
12. Nelson R. Winter S. 1977, In search of a Useful Theory of Innovation, *Research Policy*, 6 (1), 36–76
13. Dosi G. 1982, Technical Paradigms and Technical Trajectories, the Determinants and Directions of Technical Change and the Transformation of the Economy, *Research Policy*, 11, 147–162

14. Valverde S. Solé R. Bedau M. Packard N. 2007, Topology and Evolution of Technology Innovation Networks, *Santa Fe Institute Working Paper*, 06-12-054
15. Bonomi A. 2017, A Technological Model of the R&D Process and Its Implications with Scientific Research and Socio-Economic Activities, *IRCrES Working Paper*, 2/2017
16. Bonomi A. 2017, A Mathematical Toy Model of the R&D Process, How This Model May Be Useful in Studying Territorial Development, *IRCrES Working Paper* 6/2017
17. Bonomi A. 2019, The Start-Up Venture Capital Innovation System, Comparison with Industrially Financed R&D Projects System, *IRCrES Working Paper*, 2/2019
18. Cicero S. 2017, From Business Modeling to Platform Design. https://platformdesigntoolkit.com/platform-design-whitepaper/
19. Dumbleton J.H. 1986, *Management of High Technology Research and Development*, Elsevier Science Publisher, New York
20. Lane D. Maxfield R. 1995, Foresight, Complexity and Strategy, *Santa Fe Institute Working Paper*, 95-12-106

Innovation and Technology Innovation

5

5.1 GENERAL DEFINITION OF INNOVATION

Innovation is an important human activity that involves many fields of social, economic and technological nature, and that may be described in different ways. It is in fact of interest in a book about the nature of technology innovation to consider how this type of innovation is in relation with the other various types of innovations. Innovation in general, including technology innovation, may be considered itself a cumulative and historical process. It may be defined by six major characteristics such as: difficulty of prediction, in particular in calculating its scale of diffusion, asymmetric activities staggered in time, crucial role of learning times, execution and diffusion in the act of innovating, business environment conditioning the time, scale, nature and impacts of innovation and finally their interdependency [1]. In a review of the various types of innovation, it has been highlighted the current importance of organizational forms of innovation, which are the basis for the effectiveness of organizations in developing new services and products adapted to the needs of consumers. In this manner, it has been defined 15 types of innovation besides technology innovation [2], and 12 of them are involved with technological activities.

DOI: 10.1201/9781003335184-5

5.2 THE VARIOUS TYPES OF ORGANIZATIONAL INNOVATIONS

The 12 types of innovation described in the review cited previously [2] and involved with technological innovation are presented and discussed as follows.

5.2.1 Agile Innovation

Agility can be defined as an organization's ability to involve all the entities that make up this technological, economic, societal and cultural environment, integrating flexibility in all the services of an organization and in its relations with the outside world. It concerns all the stages of the production process, from the design to the distribution of new goods and services. Technology innovation is necessarily present in the activities for innovation of products and of production processes.

5.2.2 Digital Innovation

Digital innovation represents the digital transformation of companies, due to introducing digital advances: big data, open data, robotization, digital and additive manufacturing, digital twin for production processes but also internet of things for products. All that involves also artificial intelligence and digital learning. This innovation is of course strictly linked to development of ICT, either for hardware or software.

5.2.3 Dual Innovation

Dual innovation represents an innovation that may be either of military and civil importance and it is the result of R&D activities made either in industry or in public and private research laboratories dedicated to technological innovations. The history of technology is rich of important examples of new technologies developed for military purposes and after used for important economic application such as the case of making of the atomic bomb [3]. We have presented the importance of technologies developed without economic purposes acquiring after a great economic importance in the previous chapter concerning the scientific definition of technology.

5.2.4 Environmental Innovation

Environmental innovation includes technological, economic and social aspects including environmental policies and international agreements. Fundamentally environmental innovation is linked to three main problems that are: pollution, depletion of resources and global warming. Solution of the first two problems may be found in environmental industrial models such as the natural capitalism [4] and the circular economy [5]. These models concern technologies transforming conventional production processes into environmental processes with lower energy and raw material consumptions, elimination of pollution, and making products that have long life or that may be rented, easily repaired and recycled into valuable materials. Global warming is faced by substitution of conventional carbon technologies with carbon-free technologies producing energy, and is considered in environmental policies and international agreements. Technology innovation plays a key role in reaching environmental objectives, determining the practicability and times for reaching their planned results [6]. Environmental technological innovation will be discussed in detail in the chapter dedicated to relation between technology and the environment.

5.2.5 Frugal Innovation

Frugal innovation may be considered a process that reduces the complexity and cost of a product without degrading its quality or image and it is involved in various technological aspects. It was originally developed in emerging countries, particularly in India. From the technological point of view, it represents a resource-constrained innovation aiming to develop simple, less expensive products, thus making them more accessible to low-income consumers. In multinational companies, this type of innovation may be considered present in *reverse innovation* strategies.

5.2.6 Minor (Incremental) Innovation

For minor (incremental) innovation, it is intended generally the enhancement of existing innovation and improvements about productions and products. In technology dynamics, there is a clear definition of what is an incremental innovation of technological nature. Following its model of technology, it distinguishes simple improvements of efficiency of a technology intervening on its parameters, from a real technology innovation represented by a change of the structure of a technology [7]. That may be necessary for restoring efficiency consequently of externality effects. Following the model of technology, an incremental innovation is represented by a small distance between a preexistent technology and a new technology in the space of technologies.

5.2.7 Open Innovation

For open innovation, it is intended a collaborative innovation based on an organizational practice that emphasizes the need for the company to build its knowledge capital. It is a process starting from an internal or external technological base, followed by technology insourcing, licensing and technology spinoff. That in the frame of business models and targeting new or current markets [8]. From a technological point of view, open innovation is considered present in the activity of the structures organizing fluxes of knowledge and capitals in a territory. From the technological point of view, open innovation includes a *distributed innovation* (DI) system in which the generation of new technologies occurs, not only in R&D laboratories of a company but also through contract research with external laboratories, cooperation with other companies, buying or selling patents rights or competences, and buying or venture financing of startups [9].

5.2.8 Participatory Innovation

Participatory innovation involves the participation of employees, independently of their function, in the technological, organizational and commercial innovation process of the company. Such participatory atmosphere is very important in R&D activities in order to stimulate creativity in research [10]. A particular participatory attitude concerning technology innovation was promoted since the 1930 in the research laboratories of the Battelle Memorial Institute with the creation of the figure of the *researcher-entrepreneur*, i.e. a researcher who consider his work also from an entrepreneurial point of view contributing to generation of innovative ideas for new R&D projects [11].

5.2.9 Radical (Disruptive) Innovation

Radical, or disruptive, innovations are innovations that have a great impact from the technological or economic and social point of view. In technology dynamics, there is a clear definition of radical innovation based on the general model of technology. A radical technology innovation is represented in fact by a great change in a structure of a preexistent technology with the same purpose, and then by a large distance in the space of technologies of the new technology from the preexistent technology. In technology dynamics, a radical technology innovation is considered characterized by a low probability of success in its development, but at the same time by a high probability of great returns if it enters in use.

5.2.10 Responsible Innovation

Responsible innovation may be considered an innovation that takes account of social or environmental aspects avoiding or mitigating the negative consequences. From the technological point of view, responsible technological innovations concern especially solutions for problems of pollution, depletion of resources and global warming. That in the frame of the environmental industrial systems of natural capitalism and circular economy, and for carbon-free productions of energy.

5.2.11 Reverse Innovation

Reverse innovation is the result of imitation phenomena occurring in diffusion of new technologies and products especially in developing countries. From the technological point of view, it is based on a technical examination of products and technologies of production in order to imitate them obtaining, if possible, lower cost of production or products. Such innovation is linked to frugal innovation described previously with its technology innovation implications.

5.2.12 Strategic Innovation

Strategic innovation is a core business of a company and it is linked to technological innovations through the possible use of DI strategies or the development of radical technological innovations.

5.3 CONSIDERATIONS ABOUT INNOVATION

Concluding, most of the 15 types of innovation of organizational types, described in the review [2], are more or less linked to technology innovation. Exceptions are only the case of managerial innovation, social innovation and systemic innovation that are only remotely linked. This presentation of implication of technology innovation in general types of organizational innovations shows clearly the importance of technology in the innovative activities. However, besides these types of organizational innovations, it is necessary to consider also another new type of innovation with a technological base, in which the purposes are not specifically technological, industrial or economic but rather social, meeting exigencies of people. These innovations simplify the life, reduce costs, facilitate communications, moving and offering many types of services having sometimes disruptive effects. Examples of these innovations are the social networks, or services such as Airbnb and Uber, and e-commerce such as Amazon. It is interesting to note that such innovations do not take necessary origin from university studies, research centers, etc. but rather from people with various degrees of education, through their intuitions and power of observation. These innovations may change the way of life, and are normally based on the use of ICT, able to develop typically new services in the social, but also in the commercial and economic field, using capabilities such as big data storage and cloud computing, tools such as computers or smartphones and infrastructures such as Wi-Fi and internet. These technologies are generated typically by a combination of informatic and communication expertise with new ideas for possible applications in the social, commercial and even financial and banking activities such as in the case of fintech field. These types of innovations do not exploit normally results of scientific research although such innovations use well-established technologies that exploit phenomena discovered by science [7].

REFERENCES

1. Uzunidis D. Kasmi F. 2021, "Introduction" in Uzunidis D. Kasmi F. Adatto L. *Innovation, Economics, Engineering and Management Handbook,* Vol. 1, ISTE-Wiley, London
2. Laperche B. 2021, "X-Innovation – The Polymorphism of Innovation, Chapter 52" in Uzunidis D. Kasmi F. Adatto L. *Innovation, Economics, Engineering and Management Handbook,* Vol. 1ISTE-Wiley, London, 403–439
3. Rhodes R. 1986, *The Making of the Atomic Bomb,* Simon & Schuster, New York
4. Hawken P. Lovins A. Lovins H. 1999, *Natural Capitalism, Creating the Next Industrial Revolution,* Little, Brown and Company, Boston, MA
5. Stahel W. 2019, *The Circular Economy. A User's Guide,* Routledge, Taylor & Francis Editorial Group, London
6. Bonomi A. 2022, Technology and Environmental Policies, *IRCrES Working Paper,* 2/2022
7. Bonomi A. 2020, *Technology Dynamics: The Generation of Innovative Ideas and Their Transformation into New Technologies,* CRC Press, Taylor & Francis Editorial Group, London
8. Chesbrough H.W. 2003, *Open Innovation: The New Imperative for Creating and Profiting from Technology,* Harvard Business School Press, Boston, MA
9. Haour G. 2004, *Resolving the Innovation Paradox: Enhancing Growth in Technology Company,* Palgrave Macmillan, New York
10. Dumbleton J.H. 1986, *Management of High Technology Research and Development,* Elsevier Science Publisher, Amsterdam
11. Boehm G. Groner A. 1972, *Science in the Service of Mankind,* Lexington Books, D.C. Heath and Company, Lexington, KY

The
Complexity of
Technology

6

6.1 COMPLEXITY OF TECHNOLOGY

In the study of the nature of technology and its innovation, it appears clearly the complexity of the dynamics of technology [1] and of the process of technology innovation [2]. In the previous chapters, we have explained how the complexity of the set of physical, chemical and biological phenomena, existing in a technology, may be simplified by the science of complexity considering these phenomena in terms of a set of technological operations, making possible in this way the modeling of a technology [3]. In fact, the origin of this model may be ascribed to a discussion about the nature of technology between Brian Arthur, an economist, and Stuart Kauffman, a theoretical biologist, both fellows of the Santa Fe Institute [4]. This institute was founded in 1984 by George Cowan, former scientist at Los Alamos National Laboratories and first President of the institute, and having also many supporters, in particular Murray Gell-Mann, Nobel Prize in physics, and Kenneth Arrow, Nobel Prize in economy. The founding of the institute was the result of a workshop hold in Santa Fe on 1984, whose proceedings have been published recently [5], in combination with a recent review of the research activity carried out in this institute between 1984 and 2019 [6]. This review describes many fields of study of this science, but it does not include the field of technology considered a minor subject, and would be the aim of this book to renew the interest of this science in the field of technology and its innovation for a better understanding of this complex activity. It is not the objective and possibilities to discuss here the development of further applications of the science of complexity to the study of technology, and its scope

DOI: 10.1201/9781003335184-6

is limited to description of concepts, processes and models of the science of complexity from which are derived the model of technology and of the processes and structures of technology innovation described in technology dynamics. It shall be noted preliminary that the study of technology and its innovation, from the point of view of the science of complexity, implies the abandonment of the description of systems in terms of existence of a direct relation between cause and effects, and requires the adoption of a new not linear way of thinking. Furthermore, it is necessary to abandon the easy, but not suitable, view of existence of a possible forecasting of evolution of the complex systems such as technology, substituted by a multiplicity of scenarios and accepting the associated uncertainness. The study of complexity in this manner means to be able to integrate different types of information, recognize the hidden connections and possible evolutions and determine the factors of instability and possible useful variations.

6.2 DEFINITION OF THE VARIOUS TYPES OF SYSTEMS

From the point of view of the complexity, it is possible to define three general types of systems: the simple systems, the complicated systems and the complex systems.

6.2.1 Simple Systems

They are of simple nature and ready understandable in its functioning, for example a pendulum and a perturbation have typically a linear behavior in which we have generally an effect that is proportional to the magnitude of the perturbation.

6.2.2 Complicated Systems

These systems, not readily understandable, are deterministic, governed by established laws making possible forecasting or to have solutions based on analytic-deductive procedures. An example is the clock in which it is not possible to visualize immediately the functioning but it is possible to give its full description and the laws for its functioning. As in the case of simple systems, the complicated ones have typically a linear behavior in which a perturbation has generally an effect that is proportional to its magnitude.

6.2.3 Complex Systems

These systems are characterized by the fact that it is not possible to establish deterministic laws for their behavior and are then unpredictable. Their behavior is not linear and in certain case they may be subject of great perturbations but with little effects or, on the contrary, have very great effects caused by little perturbations. These systems are of various types and are the object of study of the science of complexity.

6.3 COMPLEXITY CONCEPTS AND PHENOMENA

The science of complexity assumes that seemingly disparate phenomena, both natural or social, can be understood using a common conceptual view, and that it is possible to elaborate metaphors, analogies and finally develop models for these phenomena [7]. Used concepts and described phenomena in the science of complexity are for example emergence, adaptation, evolvability, robustness, coevolution, learning, self-organization, networking, phase transition and feedback loop. These concepts and phenomena are associated to various processes existing in a complex system, and then in technological systems, and are described as follows.

6.3.1 Emergence

It is a concept corresponding to a process forming an ordered system from the chaos. This concept was used for example in the case of biology to explain self-organization and selection processes in biological evolution [8], and in technology similarly to explain the generation process of innovative ideas from available knowledge [1].

6.3.2 Adaptation

It is a concept corresponding to a process of modification of a system under the effect of externalities in order to maintain its fitness. In technology, it corresponds to the modification of values of technological parameters or of the technology structure to conserve the efficiency of technology under the effects of externalities.

6.3.3 Evolvability

It is a characteristic process of a system that continuously evolves with time modifying its structure under the effect of internal or external factors. This process occurs either in biological or technological evolutions and, in this last case, it occurs in organizational structures for technological innovations such as the R&D system, the SVC system and the industrial platform system [1].

6.3.4 Robustness

It is a phenomenon represented in natural systems by the resistance to disruption because of various externalities that act on the system. In artificial or technological engineered systems, robustness represents the resistance to disruption although under the effect of externalities not considered in the design of the system.

6.3.5 Coevolution

It is an evolutionary concept corresponding to a process observed for example in biology and concerning the genetic evolution of the prey-predator system. If a predator modifies its genetics improving efficiency in hunting a prey, it is observed also the formation of a genetic modification in the prey to compensate the increased efficiency of the predators. Such phenomenon observed in biological evolution has been called *Red Queen Regime* [9]. An analogous phenomenon is observed also in technological competition among firms of an industrial district or sector in which the competitivity obtained by an incremental innovation of a firm is readily eliminated by innovations obtained by the other firms. That leads to a continuous development of technological innovations of incremental type but without important economic growth or emerging of dominating firms [1]. The possibility to have a biological and technological interacting coevolution in the history of development of the human species is also a possible hypothesis discussed previously in the chapter about the nature and concept of technology evolution.

6.3.6 Learning

It is a general concept covering a phenomenon existing either in natural or artificial systems in which specific behaviors, necessary to maintain the fitness of a system in respect to various externalities, are memorized and made available when necessary. A typical learning activity in technology is observed for example in facing externality effects and consequent acquisition of knowhow during the use of a technology. That is also the case of LbyD activity leading to improvements or even new incremental technologies [1].

6.3.7 Self-Organization

It is a phenomenon linked to emergence in which chaotic elements become self-organized forming an ordered system and observed, as cited previously, in biology [8]. In technology, this phenomenon is represented by the formation of an innovative idea for a new technology based on a self-organization of preexistent technologies through a combinatory process exploiting or not exploiting new or never used phenomena discovered by science. The formation of organizational structures for technology innovation may be also considered as the result of a self-organization of fluxes of knowledge and capitals [1].

6.3.8 Networking

It is a typical behavior of elements of a system that enter connection forming a network. In the science of complexity networks are studied by specific models taking account of a phenomenon existing in real networks and called *small world effect*. That is based on the observation that are necessary in real networks only a very small number of passages to link two distant elements even in a great real network. The small world effect is at the origin of rapid communications in internet. The formation of networks of relations is typical of either biological or technological ecosystems. The networking of actors interested in technology innovation, and the existence of the small world effect, is important in the diffusion of knowledge useful for generation of innovative ideas for new technologies. Interesting examples of networking and small world effect are reported in technology dynamics study in the examples of diffusion of use of the general knowledge generated by R&D activities in the technology innovation system [1].

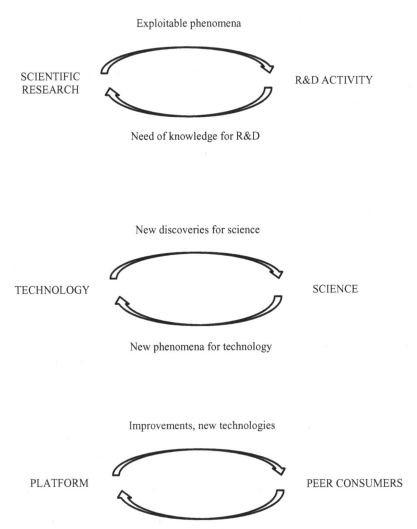

Exploitable phenomena

SCIENTIFIC RESEARCH   R&D ACTIVITY

Need of knowledge for R&D

New discoveries for science

TECHNOLOGY   SCIENCE

New phenomena for technology

Improvements, new technologies

PLATFORM   PEER CONSUMERS

Knowledge from use of technologies

FIGURE 6.1   Feedback loops in technology dynamics.

6.3.9 Feedback Loop

It is a phenomenon in which the caused effects of a system influence the factors that are the cause of the observed effects. It is formed in this way a feedback loop that determines the behavior of the system. In technology, feedback loops are observed in the generation of innovative ideas from R&D activities or during the use of technologies that are both in fact originated by previous innovative ideas. Important feedback loops exist in technology dynamics in the intertwining process between R&D activity and scientific research [1], and in the relation between science and technology [2]. In the first case, scientific results are useful for R&D activities, but R&D may trigger sometime scientific research to obtain results useful to verify possibilities of new applications. In the second case, technology is needed to make new scientific discoveries, but science supplies new discovered phenomena exploitable for new technologies. Another important feedback loop exists in industrial platforms that supply improvements and new technologies to peer consumers that return knowledge of use of technologies useful to the platform for improvements and for new technologies [1]. These three feedback loops in the field of science and technology are presented in Figure 6.1

6.4 TYPES OF COMPLEX SYSTEMS

It is possible to consider four types of complex systems: the chaotic system, the auto-organized critical system, the complex adaptive system and the network system. These systems are described as follows.

6.4.1 Chaotic System

This is a disordered system that presents neither casual phenomena statistically correlated nor adaptive behaviors. However, in many cases, observing its chaotic evolution with time, it may show some regularities. In fact, representing this system with its variables in the so-called *space of the phases*, it may be observed an approach with time of the system, often thorough a repetitive behavior, to specific values of variables constituting what it is called an *attractor basin* [10]. In another case, a chaotic system reaches a dead point ending its evolution, and it is then called a *terminated system*, or it is not observed any arrest and the system is called *progressive*. It may be observed that the fact that attractors or termination behaviors are not observed in a chaotic system, it is not a demonstration that the system does not have these behaviors because they may appear after a time greater than that used for the observations [10]. All these phenomena have been noted in the study of many chaotic systems, such as the study of the meteorological system. The knowledge associated to technological ecosystem, including the enormous number of technologies that are in use or have been used, may be considered, in a certain measure, as a chaotic system from which new technologies emerge through combinatory processes.

6.4.2 Auto-Organized Critical System

This type of systems with a chaotic behavior presents however casual phenomena that are statistically correlated. That has been observed first in 1956 by Charles Richer, an American seismologist, studying statistically the occurring of earthquakes and their energy. Richter observed that the frequency and the intensity of earthquakes, in a determined vast area for a length of time enough high, were correlated with an inverse proportional relation between the logarithm of the frequency and the logarithm of the intensity of earthquakes. This type of correlation has been observed in many other casual phenomena occurring with time, and in 1987 Per Bak, a Danish physicist, gave a demonstration of this law in an experiment using a heap of sand of critical dimension producing avalanches by addition of further sand. It was found in this case that dimensions of avalanches followed the same logarithmic law of Richter for earthquakes. He called these systems *critical auto-organized* systems [11]. In technology, the auto-organized critical systems might be represented by appearance of new technologies during technology evolution in specific fields with a major or minor degree of radicality or economic importance that might follow the Richter's law.

6.4.3 Complex Adaptive System

The *complex adaptive systems* (CAS) are the most important among the various complex systems interesting technology dynamics. This complex system is characterized by a structure consisting of a set of interactive elements that make the emergence of a certain behavior [12] or, in alternative, a system that, on the basis of its behavior in respect to its environment, carries out a specific treatment of the received information [13]. These two different views of CAS have been the object of two different models described later in the section on modeling of complex systems.

6.4.4 Network System

Network systems are the result of an activity producing networks with ordered or casual connections. From the mathematical point of view, a network may be ordered, for example the linking of atoms in a crystal, or casual in which the elements of the network are linked in a disordered way. Real networks are mostly a mixture of both presenting, as cited previously, the so-called small world effect consisting in the possibility of only a small number of connections to join even two very far nodes of the network [14].

6.5 PROCESSES IN THE COMPLEX SYSTEMS

The science of complexity has studied a certain number of processes that may occur in complex systems. Those of interest for technology are: the phase transition with autocatalytic processes and processes based on cycles.

6.5.1 Phase Transition and Autocatalysis

The process of phase transition consists in a drastic change observed in a system because of evolution of certain of its parameters. A phase transition is not a time-dependent phenomenon but dependent on changes in the structure and processes of a system. Many cases of phase transitions are known in physics concerning for example melting/solidification, vaporization/condensation processes and also transition from magnetic/not magnetic behavior as a function of temperature. Phase transitions in technology are also observed in the evolution of territories from technology stagnation or decline to technology development because of an increase of R&D activities above a critical threshold of magnitude of R&D investments [1]. A phase transition may be accompanied by autocatalytic processes of growth. Considering the relation, existing in a complex system, between the number the elements of a system and the number of their interactions, an autocatalytic process may occur when there is a sufficient high number of positive interactions among the elements. That reaching and exceeding a threshold high enough to provoke an autocatalytic process of growth. The passing of this threshold represents in fact a phase transition of the system from a subcritical stagnant situation to an autocatalytic development. This situation is represented in general by the curve reported in Figure 6.2 separating a subcritical area from the autocatalytic area. In this figure, the curve represents also the phase transition separating a subcritical from an autocatalytic situation of the complex system. In biology, it has been proposed that the formation of the first living cells occurred when a sufficiently large number of complex interactive networks were formed among molecules with a biologic potential, generating auto-sustained protometabolic networks and forming what it may be considered a phase transition from groups of simply interacting molecules to living organisms [8]. The process of phase transition, cited previously for R&D activities in a territory, has been shown studying in the model of the R&D project system [1]. In fact, making calculations about formation of successful new technologies, with a variable number of initial R&D projects in a territory, taking account also of the efficiency of the territory in exploiting the available knowledge, it may be observed an autocatalytic technology development formed when there is an enough high number of initial R&D projects started in the territory, and a sufficient efficiency in exploiting the available knowledge, constituting in fact the threshold for the autocatalytic growth of the system.

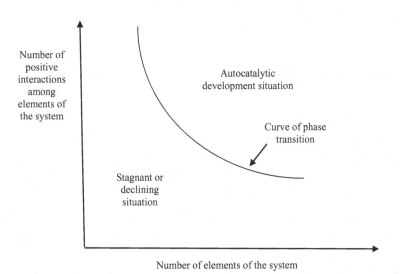

FIGURE 6.2 Autocatalysis and phase transition.

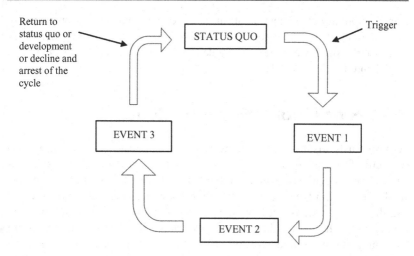

FIGURE 6.3   Example of a complex system cycle.

6.5.2 Cycles

Cycles in complex systems are phenomena characterized by a sequence of events, the last one forming anew the initial status quo event, or generating a development, or a decline and arrest of the cycle. In a certain way, cycles may be considered an evolution of feedback loops in which are introduced factors producing a connected sequence of effects in the loop. Often a cycle is started by an initial triggering factor that put it in activity. This activity may reach an equilibrium (status quo) or a continuous growth or, on the contrary, a decline until its arrest. The growth of the activity may be the consequence of an autocatalytic effect produced in the cycle, on the contrary, the decline and arrest may be produced if the magnitude of triggering is not enough effective to sustain the activity of the cycle. A schematic view of the cycle with the possible evolutions is presented in Figure 6.3. In technology dynamics, there are various important types of cycles concerning knowledge in the R&D and startups activity, and the financial cycle of the SVC system [1].

6.6 MODELS OF COMPLEX SYSTEMS

The science of complexity has developed various types of models to explain the various behaviors of the complex systems. Such models, for the transdisciplinary characteristic of this science, may find applications in many fields of science and technology. The models' interesting technology dynamics concern: the small world network model, the NK model, the fitness landscape model and the CAS models.

6.6.1 Small World Network Model

As previously discussed, real networks may be considered consisting of a mixture of ordered and casual connections among their elements and presenting the previously explained small world effect. It is then of interest to develop a mathematical model simulating the real networks presenting this effect. One of these models is for example the Watts and Strogatz model in which, in a special ordered network, some random connections are introduced between nodes of the ordered part and presenting, by running the model, a small world effect [15]. The small world effect, as previously noted, has an important role in diffusion of knowledge, for example concerning technology or innovative ideas in networks of researchers involved in a same field of R&D.

6.6.2 NK Model

The NK model is constituted by a Boolean network of N points each connected with other points with a K number of connections. The activated or deactivated state of a point will depend on the state of the K points with which it is connected through logic relations (AND, OR, NOT, etc.) chosen for the connection. A simple example of Boolean network of model NK may be formed by three points ($N = 3$) each with two connections ($K = 2$) as reported in Figure 6.4. The state of point 1 of the figure (activated or deactivated) will depend on the points 2 and 3 and by the chosen Boolean relation. For example, if the relation is AND, the point 1 will be activated only when both points 2

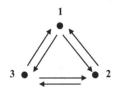

FIGURE 6.4   Schematic view of NK model with $N = 3$ and $K = 2$.

and 3 are activated. In this way, it is possible to build up quite complex networks that may assume various sets of activated or deactivated states evolving with time in a variable way, or, in certain cases, forming more or less ample zones that remain activated or deactivated or rather oscillating with time in two states. The NK model has been originally developed in physics to explain the magnetic behavior of spin-glasses. In biology the NK model has been employed by Stuart Kauffman in studying asexual biological genetic evolution [8]. The same author extended the NK model to technology substituting the action of genes with the action of technological operations, studying the dynamics of manufacturing costs in LbyD activities [3]. Another application of the NK model to technology has been also developed by Koen Frenken in 2001 [16], but considering technology as an artifact consisting of a set of components, and not as a process consisting of a set of operations as in the Kauffman's model.

6.6.3 Fitness Landscape

The fitness landscape may be derived from the NK model and represents a powerful tool in explaining the evolution of complex systems. The fitness landscape is used in particular to visualize relations between the various configurations of a system and their corresponding fitness representable in a discrete multidimensional space. If we associate to each point or configuration the scalar value of its fitness, we obtain a fitness landscape. It is possible to explain the construction of a very simple fitness landscape starting from two points of the NK model that may assume each two possible states corresponding to 1 or 0. In this case, the various configurations of the system consisting of the two points may be represented in the space of configurations by four points corresponding to the four possible combinatory strings. If we associate the scalar value of fitness for each of the four points, we obtain the fitness landscape of the system represented as example in Figure 6.5. This tool has found applications especially in theoretical biology in the study of genotypes. In technology dynamics, as previously described in the chapter about models, a fitness landscape may be used to represent the efficiency of a technology as a function of values or choices of the various parameters of each operation constituting the structure of a technology. That in what it is called a technological landscape. The case reported in Figure 6.5 corresponds in fact to a simple technology consisting of two operations each with only one parameter that may assume only the values 0 or 1. The indicated space of configuration corresponds to the technological space, and the fitness landscape to the technological landscape of this simple technology [3].

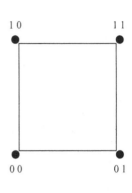
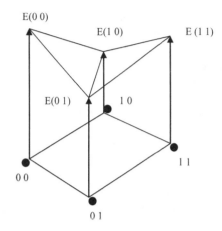

Space of configurations   Fitness landscape with E values for fitness

FIGURE 6.5   Fitness landscape of two elements of a string each with possible values: 0 and 1.

6.6.4 Complex Adaptive System Models

Adaptation phenomena are studied in the science of complexity in these types of systems, and a CAS may be modeled finding applications in many fields including technology. A first model has been developed by John Holland [12], and it consists of a set of agents that have the freedom to act in a not totally fully predictable way on the basis of own schemas, and their actions are interconnected in such a way that an action of an agent influence the actions of the other agents. More important for the study of technology is the second type of CAS model, described by Murray Gell-Mann [13]. It consists of a system that receives and treats the information acting in consequence, and it is presented schematically in Figure 6.6. The process is cyclic and starts considering an initial existence of anterior data concerning behavior, effects, etc. that the system identifies in terms of regularities and forming, through compression and simplification, a description of the forecasting behavior, and putting in this way the system in action. The operation of the system in the real world is influenced by an input of external factors that make a change of the behavior with the corresponding consequences. That has also a selective effect on the schematic structure followed by a memorization of data and experience that influence the future action of the system. This second model is of particular interest in technology as it describes the process of adaptation of a technology, under the influence of externalities, through improvements or generation of innovations during its use, as for example in LbyD activities.

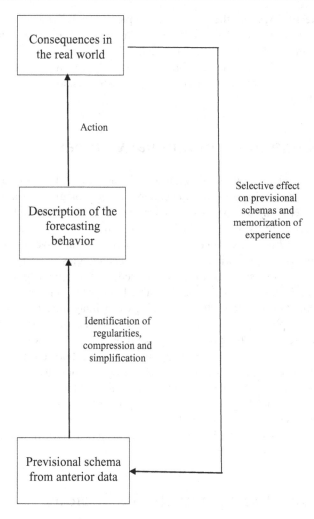

FIGURE 6.6   Gell-Mann's model of a complex adaptive system.

6.7 SCIENCE OF COMPLEXITY AND TECHNOLOGY INNOVATION

In the previous sections, we have seen how the various concepts, processes, structures and models of the science of complexity are in relation with various aspects of technology dynamics. It is of interest to see also the relation between technology and the science of complexity from the point of view of the innovation process. Therefore, considering how the concepts, structures, processes and models are in relation with the three main stages of the innovation process i.e. the generation of innovative ideas, the technological development including feasibility, development and industrialization steps and the use of technology.

6.7.1 Generation of Innovative Ideas

The process of generation of innovative ideas is based on the available knowledge concerning an enormous number of technologies that are in relation and interacting among them. This system may be considered a *chaotic system*. Partly this system might have also sometimes the behavior of an *auto-organized system* with generation of small or great waves of new more or less important innovative ideas for new technologies in certain sectors. These two types of systems are characterized by the phenomenon of *emergence* of an innovative idea from available knowledge, and consisting in a *self-organization* of preexistent technologies in imagining a possible new technology exploiting or not exploiting new or never used phenomena discovered by science. This process of generation is favored also by the existence of *networks* of people exchanging knowledge boosted by the *small world effect*. The factor controlling the emergence of innovative ideas is represented by the efficiency in exploiting the available knowledge generating more or less R&D projects or startups proposals. This efficiency, as we have seen previously, is also one of the parameters that determines the *phase transition* from a technological stagnation to an autocatalytic development in territorial innovation systems.

6.7.2 Development of the Technology

This phase of development is characterized by the presence of R&D and startups activities and for this phase are available a model of R&D based on a *knowledge cycle,* and a model of the SVC system describing the VC *financial cycle.* In Figure 6.7, we have reported as example the knowledge cycle of R&D projects or startups in technology dynamics. This phase of the innovation process is also concerned by the auto-organization of structures for innovation such the R&D system, the SVC system and the industrial platform system.

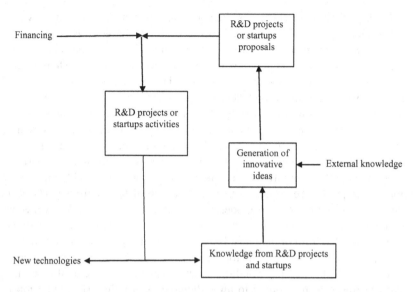

FIGURE 6.7   Knowledge cycle in R&D projects and startups.

6.7.3 Use of the Technology

During the use of technology there are numerous concepts and processes derived from the science of complexity that may characterize this activity such as: processes of *adaptation*, degree of *robustness* vs. external factors, *evolvability* with formation of new incremental innovations and *learning* through the formation of a knowhow by LbyD. An important model of the science of complexity explaining the activities during the use of a technology is a specific type of *fitness landscape* called technological landscape, presenting the efficiency of a technology as a function of its various operational conditions. A technology is normally operated in the optimal conditions represented in the landscape. Externalities may modify the form of the landscape and new optimal conditions of operation shall be found through an exploration of the new form of the landscape. Sometimes the effects of externalities cannot be eliminated by a search of new conditions in the landscape and it is necessary a technology innovation, generally of incremental type, associated to a new technological landscape. An important model of the science of complexity interesting the use of a technology, previously described, is the CAS following the Gell-Mann's view as a system treating information and acting in consequence. It may explain the way with which the use of a technology faces the influence of the various externalities. A simplified view of the Gell-Mann's model applied to the use of technology is reported in Figure 6.8. In this model, the process starts with existence of optimal conditions of operation of a technology associated with a predictive system on the behavior of the technology. That is constituted by the technological landscape and the available knowhow. Under the effects of an external factor, the used optimal conditions may be disrupted, and it is necessary to have an action to modify the operative conditions of the technology considering the landscape and accumulation of new knowhow. That occurs normally by modifying the operative conditions and possibly by introducing an incremental innovation. The new operative conditions close the CAS cycle with a possible modification of the predictive system. The interpretation of the improvements of a technology through the CAS model may explain also because a knowhow of a technology cannot be transferred completely by simply oral explanations or in the written form of manuals. In fact, a technology cannot be considered a simple deterministic system for which it is possible to give a complete description of its operative conditions. That because it is operated in a chaotic environment undergoing to unpredictable externality effects that must be considered in order to maintain its efficiency, and practically that cannot be completely included in manuals or oral descriptions, but treated as described in the CAS model. This situation well explains the necessity of LbyD in the improvement or transfer of a technology

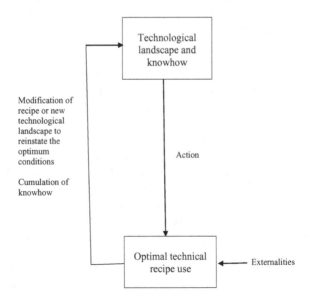

FIGURE 6.8 Technology use as complex adaptive system.

6.7.4 Feedback Loops in Technology Innovations

There is a process described by the science of complexity, the *feedback loop*, that in fact involves all the three stages of the innovation process. In the process of technology innovation, there are two phases involved in the generation

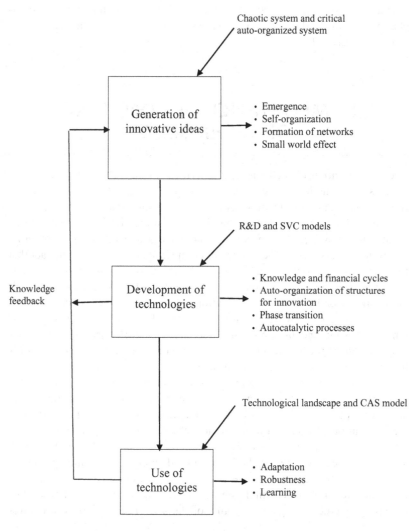

FIGURE 6.9 Science of complexity and the technology innovation process.

of knowledge. The first one is concerned with the R&D and startup activity, and it is constituted by general knowledge formed in successful or abandoned projects or startups. The second one is represented by knowledge generated during the use of a technology that may lead to incremental innovations in response to externalities or LbyD, although rarely forming radical innovations. Both sources may feed knowledge, especially in the case of R&D activities, to the stage of generation of innovative ideas that enables the starting of the following development phase. The feedback loops completed with indications of concepts, processes, structures and models derived from the science of complexity, and involved in the various phases of the innovation process, are indicated in Figure 6.9.

6.8 IMPORTANCE OF COMPLEXITY OF TECHNOLOGY

The interpretation of technological activities through concepts, structures, processes and models of the science of complexity has allowed to clear some fundamental aspects of technology and its innovation that may have applications in technology management and innovation policies [1]. Actually, there is an important aspect of processes and structures studied by technology dynamics, and linked to the science of complexity, consisting in the fact that they have been originated and evolved historically mostly as spontaneous self-organized phenomena, and not resulting by a design developed in business schools or by forecasted evolutions elaborated in the academic field, in fact both active essentially only in their study after knowing their existence. The spontaneous evolution of technological structures and processes may be considered the result of a Darwinian selection of various attempts to satisfy the various technological needs of the society. The derived technological innovation system appears consequently robust, in the sense this concept is defined by the science of complexity, assuring the generation and continuity of activities for example of the various organizational structures for innovation. Actually, the fact that technologies are the result from an innovation system formed mostly spontaneously, submitted to a Darwinian selection, constitutes one of the important advantages of technology resulting from its complexity. This overall process is characterized by the emerging of new technologies from the chaotic system of technological knowledge, an emergence occurring in the frame of the self-organization of structures for innovation such as the R&D, SVC and platform systems. On the other side, the enormous chaotic availability of technologies, and the progress in scientific discoveries, makes possible an enormous number of potential new technologies through the combinatory process of their formation. The combinatory nature of technology innovations explains well the exponential increase of available technologies observed in the evolution of technology. All that allows to conclude that technology may constitute a great potential solution for economic or environmental problems, and not just as a source of problems, and that a sustainable technologic growth is possible by a right development and use of technologies.

REFERENCES

1. Bonomi A. 2020, *Technology Dynamics: The Generation of Innovative Ideas and Their Transformation into New Technologies*, CRC Press, Taylor & Francis Editorial Group, London
2. Bonomi A. 2021, On Search of a General Model of Technology Innovation, *IRCrES Working Paper*, 4/2021
3. Auerswald P. Kauffman S. Lobo J. Shell K. 2000, The Production Recipe Approach to Modeling Technology Innovation: An Application to Learning by Doing, *Journal of Economic Dynamics and Control*, 24, 389–450
4. Waldrop M. 1992, *Complexity, the Emerging Science at the Edge of Order and Chaos*, Simon & Schuster, New York
5. Pines D. (editor) 2019, *Emerging Syntheses in Science, Proceedings of the Founding Workshops of the Santa Fe Institute*, The Santa Fe Institute Press, Santa Fe
6. Krakauer D.C. (editor) 2019, *World Hidden in Plain Sight*, The Santa Fe Institute Press, Santa Fe
7. Gray D. Macready M. 2019, "Metaphors, Ladder of Innovation" in Krakauer D. *World Hidden in Plain Sight*, Santa Fe Institute Press, Santa Fe, 129–145
8. Kauffman S. 1993, *The Origin of Order. Self-Organization and Selection in Evolution*, Oxford University Press, New York
9. Van Valen L. 1973: A New Evolutionary Law, *Evolutionary Theory*, 1, 1–30.
10. Gleick J. 1987, *Chaos*, Viking Penguin Inc., New York
11. Bak P. Tang C. Wiesenfeld K. 1987, Self-Organized Criticality: An Explanation of 1/f Noise, *Physical Review A*, 38, 364–374.
12. Holland J. 2019, "Complex Adaptive Systems: A Primer" in Krakauer D. *World Hidden in Plain Sight*, Santa Fe Institute Press, Santa Fe, 1–7.
13. Gell-Mann M. 1994, *The Quark and the Jaguar: Adventures in the Simple and the Complex*, W.H. Freeman Company, New York

14. Newman M.E.J. 1999, Small Worlds: The Structure of Social Networks, *Santa Fe Institute Working Paper*, 99-12-080
15. Watts D.J. Strogatz S.H. 1998, Collective Dynamics of "small-world" Networks, *Nature*, 393, 440–442
16. Frenken K. 2001, *Understanding Product Innovation using Complex Systems Theory*, Academic Thesis, University of Amsterdam Jointly Université Pierre Mendès France, Grenoble

Technology and the Environment

7

The relation between technology and the environment is particularly important and merits to be considered specifically also in this book concerning technology innovation. In fact, technology innovation may give an important contribute to solve environmental problems such as pollution, depletion of resources and global warming. The scientific vision of technology, and suitable models of technology and its innovation, may help the understanding of possibilities and limits of development of environmental technologies, either for production of energy in order to arrest global warming, or green technologies for elimination or reduction of pollution and depletion of resources. The knowledge of dynamics of technology may give a contribute in the estimation of practicability and possibility of the reaching of environmental objectives, such as those accompanying environmental policies and industrial environmental systems, considered in detail in a previous study [1]. Furthermore, in the relation between technology and the environment, it is necessary to discuss also the precautional principle, used in environmental policies in limiting the development of new technologies. Actually, while environment is largely studied, not only from a scientific point of view, but also for its social and political aspects, technology does not have the same fundamental attention and, in our opinion, for an effective discussion of relations between technology and the environment, it is necessary to take account of the scientific vision and the general structure and processes of technology and its innovation [2]. Concluding, the aim of this chapter is to show that technology is not necessarily a problem but a possible solution, and that a sustainable technological development is possible.

DOI: 10.1201/9781003335184-7

7.1 PHYSICAL LIMITS AND DEVELOPMENT TIMES OF ENVIRONMENTAL TECHNOLOGIES

In determining technological objectives of environmental technologies, it is important to consider the existence of scientific principles, physical constants and magnitude of natural data that limit the performance of a specific technology. In the case of environmental technologies are particularly important the laws of thermodynamics, physical constants and magnitude of natural data. For example, concerning the photovoltaic production of electric energy, the amount of its production is limited by the natural magnitude of solar irradiation of photovoltaic cells, and by the theoretical maximum efficiency by the photovoltaic material in transforming solar energy into electric energy. On the other side, the second law of thermodynamics limits the efficiency in transformation of thermal energy into mechanical or electric energy, and the concept of entropy, derived by this law, limits, for example, the possibility of recycling waste, that because the needs of energy for recycling increase rapidly with the dispersion or dilution of materials that should be recovered and transformed in usable virgin products. On the other side, technology dynamics considers that the probability of success of development of a new technology decreases with the increase of its radicality, but that developed radical technologies have a great impact and success in their use [2]. The consequence of these observations about the success of development of new technologies is the following: if we consider for example the development of a radical environmental technology, it is advisable to finance numerous various R&D feasibility projects with different innovative ideas with the same purpose. In this case it is possible to increase the probabiliy to have the development of a successful technology. Furthermore, it shall be taken in account that the possible scientific and technical criteria for selection of initial projects for radical innovations are of scarce importance. That because of the high uncertainness, characterizing the possible results of these projects and, in the case of radical green technologies, criteria based on a high environmental potential may be considered more interesting than those concerning an uncertain probability of technological success. About estimation of development times, it shall be considered that the development of a new technology is composed by a sequence of R&D projects concerning the feasibility and development steps. At the end of each project, the results are evaluated in order to take a decision to continue or to arrest the development. Furthermore, it shall be considered that it is not really possible to plan and estimate accurately the duration of next projects without knowing the results of the previous projects. All that limits the estimation of the possible duration of a development, and the existence of a great uncertainness especially in the case of development of radical technologies. Another aspect concerns the possibility to reduce the development time by increasing the availability of financing. That is possible, however, cost project management shows that this action generally also increases exponentially the cost of development becoming finally unbearable, and that the possible reduction of duration is limited by the necessary sequence of projects that shall be made for the development.

7.2 THE PRECAUTIONAL PRINCIPLE AND THE QUESTION OF RISK

The precautional principle has found a great importance in the relation between technology and the environment, and it has been largely adopted in environmental policies concerning the not-use of a new technology for which it cannot be excluded future dangers. It has been included in the Rio Declaration on Environment and Development of 1992 that formulates the principle in this way: *in order to protect the environment, the precautionary approach shall be widely applied by States according to their capabilities. Where there are threats of serious or irreversible damage, lack of full scientific certainty shall not be used as a reason for postponing cost-effective measures to prevent environmental degradation.* The precautional principle has been originally proposed by Hans Jonas [3], a German philosopher previously cited in the chapter about the nature of technology. With this principle, he contested the utopic believing, typical of the modern Western civilization, in a technology able to solve all the problems it creates. He considered that it is not correct to look only at the past or present consequences of our actions, but it is necessary also to consider the consequences in a far future that are outside of a possibility of reparation. The technology, undissociated from science, with the uncertainness of its consequences in the far future, poses questions of ethics for the humanity. In fact, if it is not possible to know the far future of our actions on the nature and the mankind, it is necessary to face this unknown by another form of anticipation such as a precautional principle. Nevertheless, the attitude of Hans Jonas does not contain in fact any disapproval to science and technology, as it would not be possible to build up a system more respectful to the environment without a scientific and a suitable technical effort [4]. Although being a very reasonable principle, it does not contain any real indication about the conditions it could be applied in practice becoming consequently the source of various different interpretations. On the other side, this principle could be overturned with the same type of arguments affirming that the *not-use of a technology* might leave consequence in the far future, outside the possibility of reparation because of the aleatory behavior of the nature generating not frequent but extremely dangerous events, such as the recent pandemic diffusion of COVID-19, for which a technology instead would had been able to supply a solution. In fact, technology dynamics has shown that new technologies are the result of combination of preexistent technologies. By consequence, the development and use of a new technology, able to face a dangerous event, might be impossible if preexistent technologies, necessary for the combination, did not have had the necessary development and use because of application of the precautional principle. In fact, all technologies are part of an ecosystem in which, by effect of the combinatory nature of innovation, the appearance or disappearance of a technology may influence the availability or unavailability of future technologies.

The problem of interpretation and use of the precautional principle is present in many directives derived by the Rio declaration about this principle in 1992. Actually, many of these directives may be criticized from a methodological point of view [1]. In fact, these directives leave undetermined the degree of scientific evidences necessary for a sanitary or environmental risk to be declared identified, and how much scientific evidence should lack to consider that a phenomenon or a human activity could be declared harmless, taking account that science cannot by principle demonstrate the complete absence of effects but only their presence. An example, well illustrating the problems of identification of potential risks of a technology in the application of the principle, is the case of presumed dangers of microwaves used for smartphones and described as follows.

Microwaves are electromagnetic waves and their behavior is well-known in physics. All electromagnetic waves are formed by a certain number of packets of energy called photons with levels of energy increasing with the frequency of the wave. The interaction of electromagnetic waves with matter is possible only if the energy of photons is sufficiently high to provoke the interaction, otherwise they will go through the matter without effects independently of their number or intensity of the wave. For example, gamma, X and UV waves, with their high frequencies, are able to break molecular bonds making damages. Infrared waves have lower frequencies and they can only interact with molecular movements resulting in a heating effect. Microwaves have frequencies lower than infrared and they have only a weak heating effect, decreasing with their frequency. That means that dangerous effects of microwaves on a human body will be possible only if it is present a low energy molecular process that may be activated by microwaves producing damage.

Science neither can indicate any possible process of this type nor can demonstrate absolutely its inexistence, leaving uncertain the application of the directives cited previously. In this case, a verification of the effects might be done through correlations in statistical epidemiologic studies that, however, are not equivalent to highly controllable laboratory experiments. In fact, the correlated effect might be the result of other causes or even by unknown factors, and only a scientific knowledge of the process at the origin of the correlation enables to consider epidemiological results as a full scientific demonstration. Actually, some statistical studies on microwave effects have shown the possible existence, not scientifically demonstrated, of a very low number of caused fatalities. However, this number of fatalities, not scientifically proved being caused by microwaves, is in fact much lower than the number of saved lives by smartphones with their alerting possibilities.

Concluding it may be argued whether a realistic alternative of application of the precautional principle might be in fact a risk evaluation that takes account not only of possible dangers of use of a technology but also the possible dangers of its not use. Actually, in more recent times, it has been recognized the difficulty of application of the precautional principle for its rigidity leading to the generation of a paralyzing effect. The UNO has elaborated in 2017 a resolution about "Objectives for a Sustainable Development", without citing the precautional principle, and using a risk-based approach or probability-based approach, alternative to the precautional principle. The advantages of this approach concern a more equilibrated evaluation of risks of danger considering not only the use but also the not-use of a technology.

7.3 TECHNOLOGY AND THE ENVIRONMENTAL PROBLEMS

In the relation of technology with the environment, there are three main environmental problems linked to technology that are: pollution, depletion of resources and global warming. Their historical acquaintance and description are the following.

7.3.1 Pollution

The problem of pollution has accompanied in fact all the industrial development of countries but the awakening of environmental actions concerning this problem may be dated back with the publication of a book of Rachel Carson in 1962 entitled "Silent Spring" [5] documenting the adverse environmental effects caused by the indiscriminate use of pesticides, and accusing the chemical industry of spreading disinformation. The problem of pollution of the environment concerns not only industry and the production of energy, but also other activities, such agriculture, transportation and domestic activities for heating and use of appliances. In transportation, we assist to the shift of combustion motors to electric motors, and that implies a reduction of pollution but an increase of needs of electrical energy in substitution of gasoline, and then the question to use a technology of production of electric energy respecting the environment and avoiding greenhouse gas emissions. A certain support to the solution of pollution problems may be found in environmental industrial systems such as the natural capitalism and the circular economy that will be treated further in the frame of the environmental industrial models.

7.3.2 Depletion of Resources

The problem of depletion of resources has been put to an international attention by publication in 1972 of a report of the Club of Rome entitled "The Limits of Growth" [6]. In this book, it was presented a scenario, derived from a global model of development, about the consequences of depletion of resources on growth. Written by a group of experts sponsored by the Club of Rome, it signaled the danger of an excessive consumption of resources. Although the authors wrote that calculations were made considering the knowledge at that time about the estimated amounts of resources, which might be discovered greater in the future, the book was perceived sometimes as a lugubrious prophecy considering a forecasting what in the reality was only a possible scenario. That diffused in the economic and political milieu a refusal to consider the problem of depletion of resources. Actually, one of the main problems raised by the book was linked to the production of energy, presently removed apparently by the discovery of new great deposits of combustibles such as coal, shale oil and natural gas, but with their use associated to problems of global warming. Actually, the problem of depletion of resources, besides the questions on combustibles, persists in other numerous cases, for niche materials necessary for new technologies. That is the case for example of rare earths for electronic applications, cobalt for high-performance magnetic materials, lithium for automotive batteries and many others. Such problems might be solved by reducing consumptions, finding alternative technologies not using such materials, or looking to extraction of these materials from very low concentrated ore or waste requiring however high consumptions of energy. In fact, both approaches of natural capitalism and in particular of circular economy, discussed forward about environmental industrial models, offer solutions to decrease sensibly, but not eliminating completely the depletion of resources.

7.3.3 Global Warming

The complex question of global warming had found a general attention since the Earth Summit, organized by UNO at Rio de Janeiro in 1992, with the creation of an international environmental treaty called United Nations Framework Convention on Climate Change (UNFCCC), followed by an agreement in 1997, called the Kyoto Protocol, establishing policies for limiting emission of greenhouse gas in the atmosphere. This protocol was signed by the 36 countries participating at this first summit. The problem of global warming is linked to the existence in nature of an important cycle based on the carbon element. In its reduced form of carbon compounds (combustibles), it may be oxidized (burned) producing energy including that (glucose) used by living species for their vital activity. All that occurs producing essentially carbon dioxide (CO_2). This last compound is transformed anew in reduced carbon compounds exploiting solar energy by natural photosynthesis, closing in this way the cycle. Photosynthesis is assured on the earth in particular by existence of plants, algae and cyanobacteria. Presently, the anthropic activity of production of energy is in a great measure assured by the use of fossil carbon in form of coal, oil and natural gas, and then by exploiting the chemical energy accumulated in the past in form of fossil carbon in period in which the photosynthesis process was boosted by high temperature and great availability of CO_2. The present emitted CO_2 of anthropic origin, added to natural CO_2 emitted by volcanic activities, outside the possibility of photosynthetic recycling, increases the concentration of this gas in the atmosphere. In this way, the equilibrium of the carbon cycle is broken cumulating an excess of CO_2 that is the cause of the greenhouse effect and then of a global warming. Although we do not know with exactitude the irregular amount of CO_2 emitted by volcanic activities, the amount of anthropic emission, cumulated since the beginning of industrial activities, is considered at the origin of the observed rapid increase of CO_2 concentration in the atmosphere. Discussing the consequent global warming, it is however necessary to consider that we are in a cyclic phase of global warming already observed in the past. However, it has been noted in the present phase that there is a rapid increase of temperature, much higher than that observed in past phases, and that may be attributed with high probability to the anthropic activity existing since the last century with great emission of CO_2 by conventional processes of production of energy and transport. The past speed of increase of temperature was lower than in the present time, and it was easy for the nature to adapt. Now this change is rapid, raising quickly a lot of problems in the ecosystem. The rapid global warming may for example produce highly energetic atmospheric phenomena, and melting of ice in the polar regions with rapid increase of the sea level and invasion with water of coastal cities. The problem of global warming will be discussed further considering the various environmental technologies producing energy, and their technological implications with environmental policies.

7.4 INDUSTRIAL ENVIRONMENTAL MODELS

It is recognized that use of technologies may be the cause of environmental damages and diseconomies of various nature and that have been described for example in a discussion about the decreasing returns of technology [7]. However, it should be noted that, from the technological point of view, it cannot be excluded that conventional technologies, with dangerous effects on the environment, could not be substituted by environmental technologies that are not necessarily more expensive because they might reduce energy consumption, raw material consumption, and waste production, as well as avoiding at the same time costs of control and elimination of pollution. That would be even economically more favorable if there is the formation of an *environmental technology ecosystem* in which environmental technologies, substituting conventional technologies, interact in a synergic way. In fact, technologies operate in an ecosystem in which each technology is influenced by other technologies through externalities of economic and physical nature. Actually, an environmental technology, included in a conventional technological ecosystem, may present diseconomies that could disappear with the formation of an environmental technology ecosystem with synergic relations with other environmental technologies. A situation in which economy of input and output of the production processes are economically and environmentally more favorable. Furthermore, in an environmental technology, ecosystem is easily possible the existence of the same operations in different technologies, that may favorize the development and use of new green technologies exploiting knowledge of operations of already existing technologies. The possibility of a technological evolution compatible with both economic and environmental aspects may be discussed considering two types of approaches. The first one, called

Natural Capitalism [8], proposes that natural resources shall be considered a capital analogous to other types of capitals involved in the economic activity, and giving suggestions about the changes that might be made to optimize the use of the natural capital, contributing in this way to the solution of environmental problems such as pollution, depletion of resources and global warming. The second one, called *Circular Economy* [9], considers the potentiality of a new industrial system that integrates production, products and their recycling in order to increase the duration of products, their possible substitution with services and an effective recycling of waste in order to obtain as much as possible virgin materials that can be reused for productions eliminating the depletion of resources. These two environmental industrial models are presented in detail as follows.

7.4.1 Natural Capitalism

The natural capitalism approach to environmental problems and their economic impact, had its origin in the USA, and presented in detail in a book written by Paul Hawken and other two co-authors entitled *Natural Capitalism* [8]. This book would show that valid technological solutions, some of those already existent, may defend and valorize the environment with an increase of natural resources and not with their destruction. *The objective of natural capitalism is a transformation of the actual socio-economic system into a system compatible with the environment.* In fact, it considers that the economic-productive system may survive only within the limits of the global ecosystem. The authors think that considering the environment, economy, social policies in competition is a prejudice, and the best solution shall not be found in a compromise that assures an improbable equilibrium among them, but by an integrated solution unifying these factors at all levels. In the book are considered various types of capitals, not only the human capital constituted by labor, intellectual property, culture and organization, financial capital constituted by money, investments and monetary instruments and asset capital constituted by equipment and buildings, but also a natural capital constituted by raw materials and living systems. The natural capital is in fact the result of a complex activity of many living ecosystems that interact with natural phenomena. The substitution of natural capital with other types of capital is possible through technology but has its limits linked to factors that auto-regulate the conditions of the atmosphere, oceans, the cycle of waters, photosynthesis, cycle of natural or anthropic wastes, protection from cosmic rays, all that making possible the life on the earth. The present industrial and economic system uses the first three types of capitals for the transformation of natural capital into economic and social goods. In the traditional capitalism, it is accepted to consider the environment, but this attention is equilibrated by the necessity of an economic growth and the maintaining of high standard of life. The history of the continuous increase of population indicates that at the beginning, there was an abundance of natural capital in terms of energy, raw materials, etc. and scarcity of human resources, now, on the contrary, we have a scarcity of natural capital and abundance of human resources. Using the same economic logic of the industrial system, it is necessary a compensation by making resources more productive improving the efficiency with which the natural capital is used. In fact, the environment represents the shell that contains, provides and sustains the whole economic and industrial system, and production technologies shall take in consideration all types of capitals including the natural one. There are four strategies suggested by natural capitalism:

- the radical increase of productivity of resources;
- bio-imitation in the production processes;
- an economy of fluxes and services based, instead on goods and purchase, on quality, usefulness and performance, changing a mentality of buying as measure of affluence and wellness; and
- making investments in natural capital enabling as much as possible services and resources.

The increase of productivity of resources means the obtention of a product with less materials and energy, improving indirectly the quality of life, showing that environment and business are not in contrast or even in conflict but on the contrary compatible in a more efficient system. Bio-imitation means a logic of production similar to that of biological systems substituting heavy structures and combustions with minimal inputs, lower process temperatures and pressures and reactions of enzymatic type (new catalysts). New fluxes and services mean that a product with a certain function, instead of be purchased, is offered as service optimizing its use and maintenance, entering a more efficient flux for recycling in a strategy similar to that of the circular economy that will be discussed later. Investments in natural capital are necessary to maintain a constant and suitable supply of services to a population in constant increase, and shall be accompanied by technological developments necessary to use efficiently such capital. All that leads to a dematerialization of products and production systems, a decrease of consumption of energy for the production, an increased productivity of resources, an efficient closure of the cycle of resources, limited or even eliminated pollution and toxicity, and finally longer life cycles of products. In Figure 7.1, we have schematically reported how natural capitalism considers the transformation of a conventional process of production into an environmental process of production following the suggested strategies. It may be observed how the transformation of a conventional

CONVENTIONAL PRODUCTION

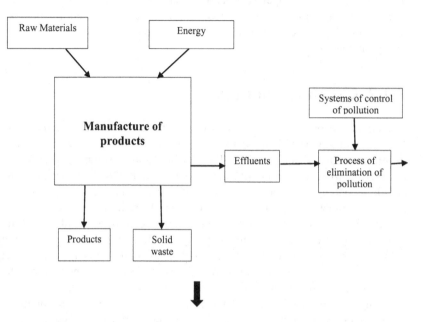

ENVIRONMENTAL PRODUCTION

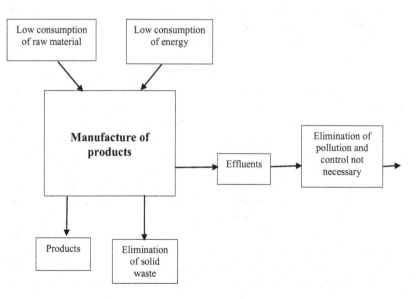

FIGURE 7.1   Conventional production process to environmental production process.

process of production into an environmental process might present possibly many economic advantages such as lower energy consumptions, lower raw material consumption, less waste production and treatment, elimination or reduction of gaseous emissions and effluents to be treated and cost of necessary pollution controls.

7.4.2 Circular Economy

The circular economy approach had origin in Europe at the beginning of 1980, in particular with the publication made in 1982 by the European Commission of a report entitled "The Potential for Substituting Manpower for Energy" written by Walter Stahel and Genevieve Reday. The basic ideas of circular economy have been condensed and updated in a recent book written by Walter Stahel [9] considered in this chapter. A preliminary observation is that this book treats in large measure policies, social aspects and industrial strategies of the circular economy and only in a limited generic way its technological implications. Following the circular economy model, presently it exists a linear industrial economy that is linked to a circular economy through the point of sale of products or services. In the model, the linear industrial economy is represented by industries involved in extraction and exploitation of resources, manufacturing and reaching the point of sale for products and services. The circular industrial economy starts from the point of sale and involves the use of products, repairing, return to the producer for repairing or upgrading for reusing and recycling of waste materials recovering materials as virgin materials that may be used anew for manufacturing. *The objective of circular economy is the development of circular conditions in such extended manner to incorporate the linear industrial economy into the circular economy*, and exploiting in this way all its full economic, social and environmental advantages, and forming what it is called a *mature circular economy*. Both the schematic views of the linear industrial economy connected with the circular industrial economy and the transformation into a mature circular economy are reported in Figure 7.2. Circular industrial economy differs from linear industrial economy because its objective is to maintain value not to create added value. The circular economy employs local small-scale processes operated by craftsmen and availability of do-it-yourself and repairing centers in order to extend the service-life of manufactured objects, as well of regional industrial remanufacturing workshops and factories to achieve the same objectives. Sustainability and circular industrial economy are two faces of the same coin. In fact, they maintain existing resources to fulfill market needs instead of relying on new materials and energy resources. By extending the service life of goods circular industrial economy employs labor-intensive activities replacing the production of new goods substituting manpower for energy, considering that human capital is not only a renewable resource but also improvable through education and training. A mature circular industrial economy will integrate the linear industrial economy into the single loop with the use-value replacing the exchange-value as central economic value substantially decreasing the greenhouse effect and increasing the number of jobs. It is possible to distinguish two eras of development of the circular industrial economy: the era of R reuse and service-life extension of goods, and the era of D recovering of waste and transforming them in pure virgin materials. The era of R is controlled by owners-users and it can appear inhomogeneous

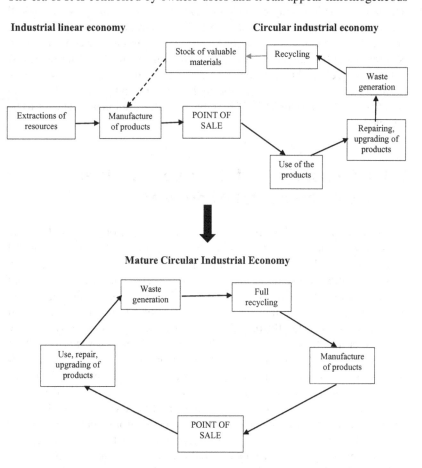

FIGURE 7.2   Industrial linear economy to mature circular industrial economy.

because the stocks of goods in use are dispersed geographically with a high diversity, and R activities may be local for tailor made objects or regional for manufactured of mass-produced goods. The era of D is controlled by economic actors for end of service-life objects and needs material and technology innovation to sort high volume and low value of waste materials turning it in recyclable goods. In a mature circular industrial economy, solutions of era of R should be preferred over solutions of era of D. The era of R aims to maintain infrastructures, buildings, equipment, vehicles, goods and other manufactured objects at the highest utility and use value at all times. The era of D needs actions to recover materials at the highest quality, pure as virgin materials, and includes for example technologies such as depolymerization, dealloying, devulcanizing noting however that these technologies are not still available and require R&D efforts. In a modern circular industrial economy, production becomes segment of the loop producing innovative components. It is recognized that the sector of circular industrial economy, with the biggest potential of technical innovation and research, is in the era of D, and concerns recycling of waste materials in the highest utility and value, and opening new fields in the development of reusable manufactured materials easy to be recycled.

7.4.3 Comparison of the Two Industrial Environmental Approaches

The birth of ideas about circular economy preceded of more of 15 years those about natural capitalism, and it has been included in environmental policies especially of EU, while natural capitalism has not obtained particular attention. Present circular economy policies concern mainly social aspects and circular industrial strategies, while technologies, although being an essential part of the industrial system, are considered only in a general way assuming practically that they are available for the purposes of the policies, or may be obtained simply by making sufficient R&D efforts. This last consideration in fact questionable following the knowledge of technology dynamics. On the contrary, the natural capitalism enters fully in the relation between technology and the environment, giving specific directives about the environmental technologies that shall be developed. At the same time, it includes many aspects favorable to the environment, considered in the circular economy, but not the full recycling objective of a mature circular economy. Both environmental industrial systems have scientific and technological limits in the possibility to reach their objectives, and that will be discussed as follows.

7.4.3.1 Limits to the Natural Capitalism

The interest and originality of the natural capitalism approach is in its search of an integration of technologies and environment, and not simply a compromise between economic aspects of technologies and environmental exigencies, and that looking for technologies that at the same time are more economic and respectful of the environment. The major limitation of this approach is in the ambitious objective to substitute all conventional technologies with really valid environmental technologies building up, if possible, an entire environmental technological ecosystem. That makes necessary important efforts in R&D but accompanied by uncertainty on duration of efforts and on the reaching of the objectives. Another limit that may be considered is of thermodynamic nature, and corresponds to the minimum but necessary need of energy to maintain viable the environmental technology ecosystem. This amount of energy consumptions may be a limit for some considered advanced objectives. Comparing with circular industrial economy, the objectives of natural capitalism seem, however, to be easier to attain, in respect to the objective of a full realization of a mature circular industrial economy.

7.4.3.2 Limits to the Circular Economy

It should be noted, at first, that the circular industrial economy represents a real novel approach to industrial production supporting an integration in making and using products, and considering of great importance the recycling of waste obtaining materials that may be used anew for manufacturing. All that represents a favorable action about reduction of pollution and of depletion of resources. Actually, the major limitation that may be considered for this model involves principally its objective of full inclusion of the linear industrial economy into the industrial circular economy transformed simply as a part of the cycle. That requires the full recycling of wastes, transformed in virgin products, usable by the manufacturing step, eliminating practically the consumption of resources. There are two main critics to this objective. The first is of thermodynamic nature and concerns the needs of energy to collect and transform waste in virgin materials that, for entropic reasons, becomes very high as the dispersion and dilution of materials to be recovered increase. These thermodynamic limits have been in fact already cited in previous studies on circular economy [10]. On the other side, it is doubtful that there are reasonable technologies able to separate in certain case virgin components, for example in the case of alloys or plastics, usable for the fabrication of the same products as considered in a mature circular economy. Furthermore, there are other two limits of the circular economy that influence indirectly but negatively the necessary technological efforts to face economic and environmental problems. The first one concerns the fact that circular economy does not have any real management strategy for innovations of radical nature that makes rapidly obsolete existent products and their recycling, leading implicitly to the paradox to hinder radical innovations with their benefits in order to conserve the circularity of the economy. In fact, radical innovations would generate a great quantity, not only of obsolete products but also of obsolete equipment for repairing, reusing and recycling worsening the operability of the circularity. Actually, a mature circular economy consists in the fabrication of a product, its use and transformation in wastes that are recycled forming virgin materials necessary to make the *same* product. A mature circular economy is then a closed technological ecosystem that may be disrupted by the appearing of innovations or natural events, for example concerning availability of energy. That without having the reserve of technologies, possibly existing outside its ecosystem, to be combined to obtain new technologies able to face the events. The second one concerns the fact that circular economy is focused on R&D efforts to change production processes in terms of reuse and recyclability of products, and not to the transformation of conventional production processes into environmental processes generating an environmental technological ecosystem, as considered in the natural capitalism system. For example, circular economy tends to renounce to bio-technologies, considered in natural capitalism, because of their linear nature not integrable in a mature cycle [9]. That may influence in fact negatively the solution of the problems of industrial pollution and global warming.

Taking account of the observations made on these two industrial environmental models, it might appear that the most efficient model could be a combination of these two systems in which the transformation of conventional processes into environmental processes is accompanied by development of a circular economy, but not necessarily searching a full transformation into its mature version, and accepting new radical technologies, even if they interfere with the recyclability of products. Finally, it shall be noted that neither the natural capitalism nor the circular economy enters in specific discussions about the necessary technological transformation of conventional into environmental productions of energy for a solution of global warming. This point will be discussed in the next section about the relation between technology and the problem of global warming.

7.5 TECHNOLOGY AND THE PROBLEM OF GLOBAL WARMING

The emission of greenhouse gas is presently important especially for industrial and agricultural activities, followed by transportation and then by domestic emissions. Concerning environmental technologies producing energy, it is necessary to distinguish between technologies presently usable and those under development. Available environmental technologies for production of energy are based on: nuclear fission, hydroelectric, biomass, wind and photovoltaic technologies, while environmental technologies under development are based on: nuclear fusion, production of combustibles by solar thermal energy or artificial photosynthesis using water and atmospheric CO_2. A further possibility is also the use of the hydrogen cycle for the production of energy, with hydrogen obtained with environmental technologies from water without greenhouse gas emissions, and producing thermal or electric energy returning to water. For technologies under development, we may add also those catching and storing CO_2 emissions. A more detailed description of these technologies with their possibilities and limits may be found in a previous study concerning technology and environmental policies [1]. Considering the possibility of a full substitution of conventional technologies for the total arrest of global warming, only the nuclear fission is presently available and experienced, assuring a continuous production without limits of power or localization and need of energy storage. Its further diffusion is, however, contrasted by its considered potential dangers following the major accident of Chernobyl and Fukushima, although improvements or even radical changes of this technology may reduce the risk of accidents. Hydroelectric and biomass productions are very effective technologies but they cannot be considered suitable for a complete substitution of conventional technologies because of limits on available sites for hydroelectric production and raw material for biomass production. Concerning wind and photovoltaic energy, at the moment they do not appear suitable for complete substitution of conventional technologies. In fact, they have irregular intermittent production with peaks not corresponding to peaks of energy consumption, and absence of production during the night for photovoltaic cells. That necessitates the storage of electric energy, a technology, necessarily decreasing the efficiency of energy availability, and not still completely developed for a large efficient capacity of storage, indispensable for a complete global substitution of conventional technologies. Photovoltaic energy is on the contrary well suitable for domestic and buildings consumptions. Photovoltaic production of energy for industrial or other generic uses needs very large surfaces of photovoltaic panels, necessary for high power productions. That may be a limit especially in industrialized countries as in Europe. About environmental technologies under development, energy from nuclear fusion appears the more advantageous as it does not have limits of power and localization and with less potential dangers than nuclear fission production. Considering then the technologies of catching and storing CO_2, they may have interest for elimination of CO_2

emissions of industrial plants but they appear unsuitable for a reduction of CO_2 in the atmosphere. In fact, this gas is extremely diluted in the atmosphere and its transformation in a concentrated form suitable for its storage needs, for thermodynamic reason, great amounts of energy. Furthermore, there are thousands of billions of tons of CO_2 to be eliminated in the atmosphere, and that might require the use of an enormous number of plants operating long times to reach the objective of a reduction of CO_2 concentration. Nevertheless, considering the wind and photovoltaic technologies as more suitable for the substitution of conventional technologies, the question is not if they would not find in future solutions for the present limitations, but whether these solutions *will be available before than the increase of temperatures, because of global warming, will manifest deleterious effects in a great measure.* In fact, the arrest of greenhouse gas emission *will not decrease the deleterious effects of global warming, as these depend of the CO_2 concentration in the atmosphere, and not by the arrest of its emission.* Technologies under development present the same question and probably even long times for their availability and diffusion. This situation is worsened by a probable increase of energy demand by countries under development. In fact, without their choice of environmental technologies for their needs of energy, these countries might cancel the efforts of industrialized countries in reducing the emission of greenhouse gas [11]. All that raises a question about the present environmental strategies that do not consider sufficiently the development also of technologies that can mitigate the expected unavoidable dangerous effects of global warming [1].

7.6 TECHNOLOGY INNOVATION AND GREEN TECHNOLOGIES

Green technologies represent an important aspect of the contribution of technology to the solution of environmental problems. The technology innovation in this field may be discussed on the basis of the type of environmental problem concerned with the green technology, and taking account of the radical degree of the corresponding innovation. Following technology dynamics, it is expected that radical technologies are the more difficult to develop but also much more effective, when successful, in solving environmental problems. Finally, it may be discussed the suitability of the various organizational structures for innovation following the various types of green technologies.

7.6.1 Green Technologies for Environmental Problems

The environmental problems concerned by green technology are essentially: pollution, depletion of resources and global warming.

7.6.1.1 Technologies against Pollution and Depletion of Resources

These green technologies are particularly involved in the industrial system, and obtained transforming conventional technologies of production into environmental technologies, as indicated in Figure 7.1. These new technologies may be of both incremental and radical type. The technologies of incremental type concern improvements, obtained for example by digitalization of processes, while technologies of radical type are able to fully substitute conventional production processes with environmental ones. Technologies limiting the depletion of resources are involved especially in the circular economy approach, and concern improvement of the duration of products and their reuse by reparations and upgrading, but in particular by recycling. Most of these technologies are of incremental nature, but radical technologies may be involved especially for a more effective recycling.

7.6.1.2 Technologies against Global Warming

These technologies concern the carbon free production of energy. As previously discussed, there are technologies already available, such as nuclear fission, hydroelectric, biomass, photovoltaic and wind technologies, mainly involved in incremental innovations. However, in the case of nuclear fission technologies, there are possible also radical changes with new reactors design and new types of nuclear fuels. Concerning technologies under development, such as nuclear fusion, production of fuels by thermal solar energy or artificial photosynthesis, and the realization of the hydrogen cycle for energy production, they are all of radical nature.

7.6.2 Green Technologies and the Organizational Structures for Innovation

The development of green technologies occurs, following technology dynamics, in structures organizing fluxes of knowledge and capitals described previously that are: the industrial R&D projects system, the SVC system and the industrial platform system. Each of these organizational structures offers different conditions for the development of green technologies.

7.6.2.1 The Industrial R&D System and the Green Technologies

The industrial R&D activity has been shown particularly suitable for the development of technologies with a relatively low radical degree and then higher probability of success. It is a system typically followed by industry, well appropriated for innovation of productions and products. This system may be suitable for the development of incremental green technologies, involved in the environmental industrial system, as well as for improvements of available technologies for carbon-free production of energy. However, this important aspect concerning the relation of the degree of radicality of technology, and the more suitable organizational structure for its development, has not taken in consideration in derived policies since the Kyoto Protocol. In fact, the task of technology development has been leaved essentially to the industrial R&D system possibly with public aids. By consequence, R&D was most oriented toward basically known technologies such as solar thermal, photovoltaic and wind technologies with a low degree of radicality and lower risk of failure, instead of orienting major efforts of R&D also toward more radical technologies, such as solar thermal or artificial photosynthesis for the production of fuels or hydrogen. These radical technologies have higher risk of failure but also being possibly much more efficient for a global substitution of conventional technologies. Actually, if we would have had an agreement, since the Kyoto Protocol, not only of political nature but also of promotion and coordination of a great international project for the development of various radical technologies, now, after near 30 years of development, we would have probably already available a set of these new radical technologies for a strong reduction of carbon emission.

7.6.2.2 The SVC System and the Green Technologies

The SVC system is known to be suitable for the development of radical technologies, and VC invests in this kind of technologies with the aim to sell the developed technology with high returns. In fact, VC is known to invest in green technologies with a certain high degree of radicality and possible high returns. However, certain new radical technologies, in particular for the production of energy, do not fit well with the normal financing strategies for VC, in particular for the necessary length of the development and political involvement in the diffusion of this type of technologies. For this reason, the SVC system cannot be considered a full solution for the development of radical green technologies without public financing and international cooperation. These questions have been discussed for example in a recent report about the role of VC and the government about clean energy production [12].

7.6.2.3 The Industrial Platform System and the Green Technologies

The industrial platform system is not involved directly in the development of technologies and then of green technologies. These ones may be developed by R&D projects and startups present in the structure of the platform. In fact, the function of the platform is to increase the available knowledge, boosting improvements and development of new technologies through the relations among the various actors of the platform. However, the possible evolution of this system toward the formation of an industrial platforms network [2], with platforms supplying basic green technologies to industrial companies, would be particularly favorable in the development of an environmental industrial ecosystem based on natural capitalism and circular economy.

REFERENCES

1. Bonomi A. 2022, Technology and Environmental Policies, *IRCrES Working Paper*, 2/2022
2. Bonomi A. 2020, *Technology Dynamics: The Generation of Innovative Ideas and Their Transformation into New Technologies*, CRC Press, Taylor & Francis Editorial Group, London
3. Jonas H. 1979, *Das Prinzip Verantwortung. Versuch einer Ethik für die technologische Zivilisation*, Suhrkamp, Frankfurt/M
4. Bourg D. 1993, Hans Jonas et l'écologie, *La Recherche*, 246, 886–890
5. Carson R. 1962, *Silent Spring*, Houghton Mifflin, Boston, MA
6. Meadows D.H. Meadows D.L. Randers J. Behrens W. 1972, *The Limits to Growth*, Potomac Associated Book, Falls Church.
7. Giarini O. Loubergé H. 1978, *The Diminishing Returns of Technology: An Essay on the Crisis in Economic Growth*, Pergamon Press, Oxford UK
8. Hawken P. Lovins A. Lovins H. 1999, *Natural Capitalism, Creating the Next Industrial Revolution*, Little, Brown and Company, Boston, MA
9. Stahel W. 2019, *The Circular Economy. A User's Guide*, Routledge, Taylor & Francis Editorial Group, London
10. Korhonen J. Honkasalo A. Seppälä J. 2017, Circular Economy: The Concept and its limitations, *Ecological Economics*, 143, 37–46

11. Grubb M. Koehler J. 2002, Technical Change and Energy/Environment Modelling, 27–59, in *Technology Policy and the Environment*, Workshop Paris 21 June 2001, OECD Publications, Paris

12. Van den Heuvel M. Popp D. 2022, *The Role of Venture Capital and the Governments in Clean Energy: Lessons from the First Cleantech Bubble*, Munich Society for the Promotion of Economic Research, CESifo Working Paper No. 9684, Munich

Applications of the Models

8

The models of technology and of technology innovation may find applications in many fields. A certain number of applications have found a description in the study of technology dynamics [1], and concerned for example explanation of generation of new technologies, not only during R&D developments but also during the use of a technology, the dependence of investments in R&D and the economic growth in a territory, and the possibility of new statistical studies about the innovation process for promotion policies. We present here two new possible important applications, the first one derived from the general model of technology describing its functioning, and making possible also its optimization taking in consideration various types of technology efficiency; the second application, derived from the models of organizational structures for innovation, concerns a model of the technology innovation system of a territory.

8.1 FUNCTIONING AND OPTIMIZATION OF A TECHNOLOGY

The functioning and optimization of a technology may be obtained through its technology model, considering the structure of the technology, and taking account of the external and internal factors influencing its efficiency.

8.1.1 Model of the Technology

In a previous chapter, we have described a general model of technology fully developed in technology dynamics also with its mathematical aspect [1]. We summarize here this model considering technology as a set of technological operations representable in a structure corresponding to a graph in which the arcs, oriented with time, represent the technological

DOI: 10.1201/9781003335184-8

operations. Furthermore, each operation may be controlled by a set of parameters or instructions each characterized by a certain number of choices or values existing in a determined range. Considering the technological operations, with their parameters and their possible values or choices, it is possible, through a combinatory calculation, to determine all the configurations or recipes potentially usable for a technology, and to represent all the recipes in a multidimensional discrete space called technological space [2]. Each recipe of a technology presents a scalar value of a certain type of efficiency. It is then possible to associate this scalar value to each recipe of the technological space obtaining a technological landscape [2]. The form of the technological landscape will depend, of course, on the considered type of efficiency that may be economic, energetic, environmental, etc.

8.1.2 Technological Processes

The model of technology may explain the existence of various technological processes. The main processes influencing the efficiency of a technology are represented by the externalities and intranalities of the technology, that we have presented previously in the chapter about model of technology and its innovation, and that we recall here.

8.1.2.1 Effects of Externalities

During the use of a technology, there are many external factors that may influence its efficiency. These factors may be the change of costs, raw materials, and new regulations to be complied. Such changes modify the form of the technological landscape reducing possibly the efficiency of the used recipe. It is then necessary to search a new optimal recipe exploring the new landscape resulting by the externality effects, or even change the structure of the technology realizing an innovation normally of incremental type. The effects of externalities on the technological landscape may have a mathematical description considering the existence of a certain number of external variables influencing the technology with their parameters, values or choices, and applying the same mathematical approach used for the operations of a technology. That makes possible the calculation of a number of possible configurations of external variables influencing the efficiency of technological operations, and then changing the technological landscape [1]. However, it shall be noted that it is not possible to consider all the possible external variables influencing the technology efficiency. In fact, a technology cannot be considered a simple deterministic system for which it is possible to give a complete description of its functioning as it is operated in a chaotic environment undergoing to unpredictable externality effects.

8.1.2.2 Effect of Intranality

The intranality effect consists in the fact that changing the parameters values of an operation, in order to improve its efficiency, that may influence the efficiency of other operations of the technology [2]. Consequently, in the presence of intranality effects, the optimization of the entire technology shall be obtained by a tuning work among the various parameter values or choices of the operations of a technology. The intranality effect exists also in the elimination, substitution or addition of an operation in the structure of a technology, and that may influence the efficiency of other operations of the structure [1]. The presence of effects of intranality in a technology may be determined by considering a matrix with columns representing the operations, and rows representing operations parameters, and indicating in the matrix the existence or not of an intranality effect. It shall be also noted that a theoretical study on the technological landscape has shown that in the absence of intranalities, the technological landscape has only an optimum value of efficiency on the top of a "hill". If there are intranalities, the landscape may have clusters of optimal "peaks", and, if intranalities are very numerous, the landscape may appear rugged with numerous peaks around the same efficiency [3].

8.1.3 Optimization of Operative Conditions of a Technology

The described model of technology, with their externalities and intranality effects, may simulate the functioning of a technology that may be useful in searching the optimal operative conditions of a technology considering also the externality in which it operates. The procedure of optimization is reported schematically in Figure 8.1. In this procedure, a technology is at first modelized in terms of structure of its operations obtaining its basic model. Taking account of its operations with their parameters, values or choices, it is possible to obtain all the recipes determining its technological space. Considering the involved various types of efficiency, it is possible to obtain the corresponding technological landscapes associating the corresponding efficiency to each recipe. These landscapes may be confronted with known externality effects of variables with their parameters, values and choices, determining various externality configurations. The existence of possible intranality effects may be pointed out in a suitable matrix representation and possibly taken in account. From all these data, it is possible to optimize the operative conditions of a technology taking account of the various landscapes determined by the various types of efficiency and externalities, possibly considering also the intranality effects. In this way, it is possible to obtain the optimal conditions of functioning of a technology that take account of the various types of efficiency, and the effects of externalities on the efficiency. It shall be noted that intranality effects depend on the combinatory influences among all the technological operations and may form an enormous number of configurations. It is then useful to verify whether the possible intranality effects influence really in a sensible measure the efficiency of the technology or considering only the few important effects. It may be observed also that the use of this procedure might be impracticable in the case of complex technologies, because of the possible enormous extensions of their structures, and then difficulties to give a complete fine description of the technology model in order to take account of all the parameters and values necessary to control the entire technology. However, the application of this method of optimization is possible and useful also for single operations or group of linked operations owing to an extended technological structure. In this case, it might be necessary just to integrate the procedure with possible intranality effects that might be present with the other not considered operations of the technology. Another difficulty to be overcome in this procedure is in the determination of the technological landscapes. In the case of economic efficiency, the cost of recipes may be modelized obtaining easily by computer calculation the efficiency of all recipes. However, it shall be considered that certain recipes of the technology might satisfy optimal economic conditions but not necessarily the purpose of the technology, represented by other types of efficiency, and then the necessity to take account of that in the search of optimal conditions of operation of a technology. Considering the other than economic types of efficiency, it may be necessary to make measures or even experiments to obtain the values of efficiency of each recipe, and that might be in an unpractical high number. An acceptable number may be obtained by isolating part of the technological space, on the basis of scientific and technical knowledge of the process. In this way, it is possible to reduce the number of recipes for which it is necessary to know the efficiency on the basis of their effective influence on the optimization. When all the data about the various technological landscapes are available, it is possible to determine, by a suitable mathematical procedure, the optimal operational conditions that take account of the various landscapes with the various types of efficiency, and the various conditions of externality considered for the technology.

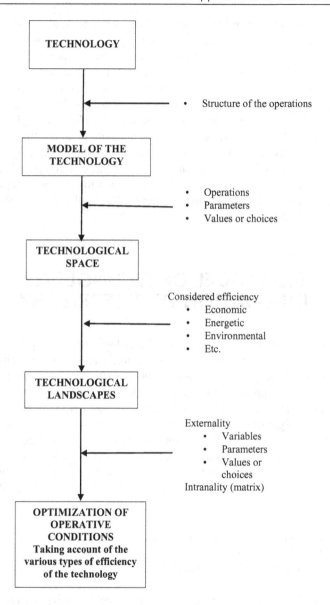

FIGURE 8.1 A schematic view of the procedure of optimization of a technology.

A simple example of utilization of this procedure has been reported as application of the model of technology reported in Appendix 1 of the previous book on technology dynamics [1]. This case concerned a technology of elimination of contamination of drinking water by lead, present on the surface of brass made faucets, by treating them in a de-leading bath. In this case, it was considered the optimization of the technology on the basis of its economic and environmental efficiency, describing also the adopted simplifications in order to reduce the number of recipes to be experimented for the determination of the technological landscape. In this case, it has been considered various scenarios of external variables, but neglecting the weak effects of intranality. In conclusion, it should be noted that this cited example of application of the model of functioning of a technology is very simple requiring only easy mathematical procedures. A computerization of the model and of the procedure of optimization, with a suitable capacity of storage of data, may extend sensibly the practical application of this model for more complex technologies.

8.2 MODEL OF TECHNOLOGY INNOVATION OF A TERRITORY

This model, derived from studies on technology dynamics [1], considers that the technology innovation system of a territory may be described taking account of the presence of the various structures organizing fluxes of knowledge and capitals that generate new technologies. That in alternative to consider technology innovation in a territory generated in the frame of institutional organizations such as universities, research centers, industrial R&D laboratories but considering that these institutions, making for example R&D activity, organize fluxes of knowledge and capitals in the same way following the structure of R&D system. The structures organizing fluxes of knowledge and capitals for technology innovation already described previously are:

- The industrial R&D projects system
- The SVC system
- The industrial platform system

This view of a technology innovation system through organizational structures of knowledge and capitals is a novelty in respect to that used in statistical studies considering investments in technology innovation, following guides such as the Frascati [4] or the Oslo [5] manuals, that see technology innovation in terms of investments and as result of R&D activities in industry and in institutional organizations, but not entering in detail how it occurs. That means also that the technology innovation of a territory does not depend simply on the presence of various institutional organizations but rather on the number of R&D projects and startups in activity and presence of industrial platforms in the territory. It shall be noted that the view of technology innovation in terms of structures organizing fluxes of knowledge and capitals is independent of industrial or economic factors, as cited previously while discussing the models of technology and of its innovation.

A model of the industrial R&D projects system has been developed in a previous study [6], and it is presented in Figure 4.3. The activity consists of a cyclic process of fluxes of knowledge and capitals generating new technologies and knowledge coming from either successful or abandoned projects. The new formed technologies may be seen also in terms of capitals that have been necessary for developing both successful and abandoned R&D projects. The generated knowledge, added to existent knowledge generated in past cycles, but reduced by a fading effect, is increased by external knowledge coming from scientific, technical or other origin, and forming in this way the total available knowledge for the generation of innovative ideas for the proposals of new R&D projects. Such generation of new ideas depends on a combinatory process of the available elements of knowledge, in accord with the idea that new technologies are formed by a combination of previously existent or abandoned technologies [7]. The formation of R&D project proposals depends on the efficiency of the territory to exploit the available knowledge. The formed R&D project proposals will be selected leading to a certain number of financed R&D projects constituting the R&D activity of the cycle. The formed new technologies need industrial capitals to be used generating possibly ROI. The industrial capital system, joined possibly with public financing, makes available R&D investment to finance new selected R&D projects closing both capital and knowledge cycle. This schematic view of the model of R&D activity was that practically existent in the technological system of a territory before the 1970s, in which startups and venture capital activities were negligible, and the industrial platform system inexistent, becoming in fact a reality only at the beginning of the XXI century. In the present situation, the technology innovation system of a territory cannot be represented only by industrial R&D activities but it is necessary to take account of existence of the SVC system and, in a certain measure, of the activity of industrial platforms. Such organizational structures differ from that of R&D projects activity and, in particular, the SVC system has a different capital strategy as it develops new technologies for their sale and not for their exploitation, as in the case of industrial capital, reinvesting part of these returns in new startups [8]. On the other side, industrial platforms supply additional knowledge to the technology innovation system of a territory. That in fact

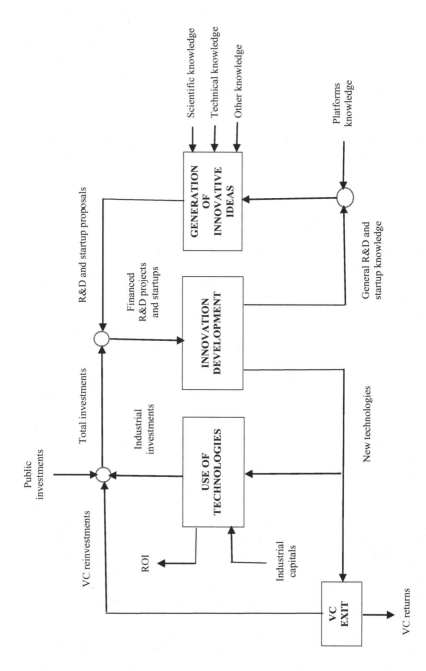

is through a continuous exchange of information concerning the use of a technology, increasing in this way the generation of innovative ideas for new technologies. It is then necessary to modify the schematic view of the fluxes of knowledge and capitals of the R&D system, as reported in Figure 4.3, in order to take account of existence of the SVC and the industrial platform systems contributing to the technology innovation system of a territory. This new representation of the fluxes of knowledge and capitals is reported in Figure 8.2. We may note that the technology innovation activity is represented not only by R&D projects but also by startups with an activity that not only includes R&D projects but also develops business models suitable for new technologies. On the other side, industrial platforms do not generate directly new technologies but through R&D projects and startups present in their structures. In this model, the flux of capitals concerns not only industrial capital financing new R&D projects but also venture capital financing new startups. Concluding, it shall be noted that this model of technology innovation considers the existence in the territory, as in the previous R&D model [6], of a distributed innovation activity consisting not only in internal industrial R&D but also in contract research, startups, and existence of corporate venture funding, industrial co-operations, technology trade [9], that in the frame of a system of open innovation [10].

8.2.1 The Mathematical Model of Technology Innovation of a Territory

As in the case of the R&D model, this model of the technology innovation of a territory may have a mathematical description. We give here only a qualitative view of the mathematical aspects of the model. A full detailed description is reported in the Appendix for readers interested in these aspects of the model. Taking account of the differences existing in respect to a territorial innovation system based only on R&D projects, it is possible to develop a mathematical model simulating the innovation activity by considering suitable modifications of the previous mathematical model of R&D activities [11]. Similarly, to the previous simulation model, the mathematical model of the technological innovation system of a territory is based on the following assumptions:

- The generation of new technologies is obtained by R&D projects and startups.

- Industrial platforms do not generate directly new technologies but by R&D projects and startups existing in their structures.
- Part of the formed new technologies becomes successful technologies contributing to the economic growth of the territory.
- The innovation activity is seen in form of cycles, each fed by a certain number of R&D projects and startups, and generating a certain number of new technologies, part of them becoming successful technologies.
- The flux of knowledge in the system is considered as formed by a certain number of information packages, generated by each successful or abandoned R&D project or startup, and by additional knowledge from platforms and from external origin.
- Proposals for new R&D projects and startups result by a combination of such packages, and followed by a selection for financing.

The mathematical model, as in the previous model [11], considers knowledge as formed by a certain number of information packages. That means that each R&D project or startup, successful or abandoned, generates an average number of information packages. The number of generated packages by startups is considered higher than that of R&D projects to also take account of generation of knowledge about business models. The total knowledge generated in a cycle by R&D projects and startups is increased by the knowledge of past cycles, although reduced by a fading effect. Furthermore, this knowledge is increased by a fraction concerning the knowledge originated by industrial platforms activities, and a fraction concerning external knowledge of scientific, technical or of other nature. The number of potential innovative ideas may be obtained by a combinatory calculation considering the available information packages. An innovative idea for new R&D projects or startup proposals results by a combination of a limited number of available information packages. In this way, it is obtained normally a great number of potential innovative ideas, many of these not valid or even absurd, and it is the ISE of the territory determining the selection of valid ideas for R&D projects or startups forming the correspondent proposals. In order to calculate the number of formed successful technologies, various selection rates concerning are considered hereafter:

- the number of R&D projects and startups proposals that are financed for a development,
- the number of R&D projects and startups effectively generating new technologies, and

- the number of successful technologies resulting from new technologies formed by R&D projects or startups.

The total list of parameters of the model is reported in Table A.1 of the Appendix. It shall be noted that the various rates concerning the selection for financing, the rate of generation of new technologies, and the rate of formation of successful technologies may be different in the case of R&D projects or startups. That is a consequence of the fact that radical innovations, typically developed by startups, have a lower probability to be fully developed and, on the other side, in the case of success, they have a higher probability to have great ROI in respect of incremental innovations, typically developed by industrial R&D projects [1]. Consequently, the rate of formation of new technologies from startups will be lower than in the case of R&D projects. On the contrary, the rate of formation of successful technologies from startups will be higher than in the case of formation from R&D projects.

The mathematical model considers the innovation activity of a territory in the form of cycles in which are active a certain number of R&D projects and startups, and taking account of the fraction of startups present in this total number. Each R&D project or startup generates an average number of information packages, higher in the case of startups also taking account of generation of business models. To the total numbers of packages generated in the cycle are added packages formed in previous cycles but reduced by a fading effect. Furthermore, fractions of information packages coming from platforms and of external origin are added. In this way, the total number of information packages available for the next cycle is obtained. Considering that a potential innovative idea may be obtained by combination of a certain number of available information packages, the number of generated potential innovative ideas is obtained by a combinatory calculation based on this number. The total number of R&D projects and startups proposals is actually a small fraction of these combinations, and depends on the ISE of the territory in exploiting the available knowledge. This rate might be different following the case of R&D projects or startup proposals. In the same manner, it is possible to calculate the number of financed startups and R&D projects considering a rate of financing that also may be different for the case of R&D projects or startups. The formation of new technologies may be calculated from these numbers of financed R&D projects and startups considering the two different rates of formation from R&D projects and startups. The number of successful technologies may be calculated from the number of technologies derived by R&D projects and startups considering the two different rates. In this way, it is possible to obtain finally the total number of successful technologies formed in a cycle.

8.2.2 Example of Calculation with the Model

The model has been used to make a calculation of the number of formed successful new technologies from an initial total number of R&D projects and startups for a series of cycles, taking account of knowledge of previous cycles reduced by a fading effect. For calculations, three scenarios have been considered. The first one with only industrial R&D activity, the second one with R&D and startup activity but with absence of industrial platforms, and the third one with R&D activity and industrial platforms but with absence of startup activity. In this way, it is possible to observe separately the variation of number of new successful technologies, in respect to only industrial R&D activity, and in the case of presence of startups or of industrial platforms. Concerning values of used parameters, some of them result from indications of studies on the rate of success of startups [8] or of R&D projects based indicatively on rate of success of patents [12] in forming successful technologies. Most of the values of the other parameters are indicative, and are based on experience in R&D and startup activities. A discussion on choice of the values of the parameters, concerning the various types of selection rates and

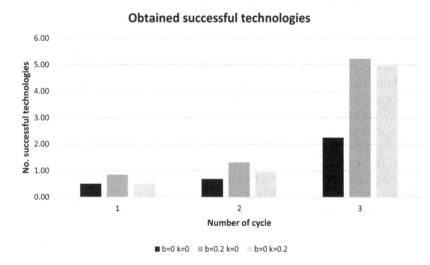

FIGURE 8.3 Number of obtained successful technologies after three cycles.

generation of proposals, is reported in the next section about the results of the calculations. For calculations, a drastic simplification has been adopted for considering that all obtained proposals of R&D projects or startups are financed, and that the total number of R&D projects and startup proposals is obtained from the amount of potential innovative ideas using a unique rate of 0.0025 equivalent to 25 valid innovative ideas, on 10.000 combinatory possibilities, becoming R&D projects or startup proposals. This value represents an average of the discussed rates used in the previous study on the R&D projects model [11]. The parameter values used for running the model are reported in Table A.2 of the Appendix. The model calculations have been made by using an Excel® sheet derived by the model represented in Figure 8.2, and introducing the mathematical equations in the suitable mathematically connected case positions. An example of sheet is reported in Figure A.1 of the Appendix. The calculations have been made starting from the initial number of R&D projects and startups equal to 100, and recording the results of each successive cycle. The results are presented in Figure 8.3 reporting the number of obtained successful technologies following three cycles for the three adopted scenarios. It is possible to observe that both the presence of startups and industrial platforms increase the number of formed successful technologies in respect to the presence of only industrial R&D projects activity. Apparently, the impact of startups is higher than that of industrial platforms; however, it shall be noted that industrial platforms are only at the beginning of their use and we do not have a real detailed knowledge of their impact on the technology innovation system to consider definitively startups more efficient than platforms. Going further in the calculation for a fourth cycle, the number of formed successful technologies becomes unreasonably high, with more than 100 successful technologies, and such result is discussed in the next section.

8.2.3 Discussion on the Model and Its Results

It should be noted, before all, that the technology innovation system of a territory is highly complex and the simulation made by this model is only a rough representation. That, joined to the possibility to use only indicative and uncertain values of parameters for its functioning, makes the model more a toy than a simulation model. However, we think that it may show some interesting aspects of the behavior of the technology innovation system of a territory that takes in consideration all the organizational structures for generation of new technologies. Furthermore, the model, in terms of fluxes of knowledge and capitals, may be applied with the same structure, not only to territories but also to one or a group of institutional or industrial organizations developing new technologies. A key critical aspect of the model concerns the generation of R&D projects and startup proposals from available knowledge. That changes following the different ISE of the territories, with their uncertainty of the value that may be used for the calculations. Another critical key element of the model is in the rates of selection of R&D projects and startups to be financed. That may be variable with time depending on adopted financial strategies and capital availability of industrial and venture capital. Actually, the number of formed new and successful technologies depends, on one side, on the combinatory possibilities of formation of innovative ideas and, on the other side, on the availability of capitals. That makes unrealistic the possibility of financing the large number of R&D projects and startups, possibly resulting from the great number of potential innovative ideas of combinatory origin, as observed by running the model with increasing the number of cycles. That is a consequence of the adopted approximation that considers the existence of financing capitals for all obtained R&D projects and startup proposals. Actually, capitals availability has always a limit, and other limits may be also the availability of human resources and facilities for R&D or startup activity. The positive impacts of presence of startups and industrial platforms in the generation of successfull technologies in a territory, that is something that has been also empirically observed for example in the Silicon Valley. The positive impact is evident in the case of industrial platforms as they supply additional knowledge favoring formation of innovative ideas, and finally successful technologies. In the case of startups that needs an explanation. In fact, the generation of successful technologies by startups is, on one side, favored by the high degree of radicality of the developed technologies and, on the other side, the high degree of radicality results in a high rate of abandonment of startup developments [8]. Actually, it is the ability of the venture capital, with its need to close positively its financial cycle, to make necessarily the right selection and support to startups in order to obtain a positive financial cycle [8], and that justifies the positive influence of the presence of startups in the technology innovation system of a territory.

In conclusion, it is not the aim and possibilities of this model to simulate the real innovation activity of a territory with quantitative results in accord with empirical evidence. Its objective is to identify the structures and processes that are present in the technology innovation system of a territory resulting from the organization of knowledge and capital fluxes. That means that the model is tentative to show how it is the functioning of the technology innovation activity in a territory, independently of industrial or economic factors. This view may be useful in technology and R&D management in determining the role of certain factors for innovation such as the conservation of past and availability of external knowledge, the necessary combinatory creativity for innovative ideas and the possible financing strategies of selection of R&D projects or startups. Other activities that may be interested by the model are the case of statistical studies that would consider in depth the technology innovation process in the elaboration of territorial policies for innovation. Finally, the model, differently by current views on the technology innovation systems, takes account of the presence of industrial platforms, supplying additional knowledge and then innovative ideas to the system.

REFERENCES

1. Bonomi A. 2020, *Technology Dynamics: The Generation of Innovative Ideas and Their Transformation Unto New Technologies*, CRC Press, Taylor & Francis Editorial Group, London
2. Auerswald P. Kauffman S. Lobo J. Shell K. 2000, The Production Recipe Approach to Modeling Technology Innovation: An Application to Learning by Doing, *Journal of Economic Dynamics and Control*, 24, 389–450
3. Kauffman S. Lobo J. Macready G.W. 2000, Optimal Search on a Technology Landscape, *Journal of Economic Behaviour and Organization*, 43, 141–166
4. OECD 2015, *Frascati Manual 2015: Guidelines for Collecting and Reporting Data of Research and Experimental Development*, The Measurement of Scientific, Technological and Innovation Activities, OECD Publishing, Paris
5. OECD Eurostat 2018, *Oslo Manual 2018, Guidelines for Collecting, Reporting and Using Data on Innovation*, The Measurement of Scientific, Technological and Innovation Activities, OECD Publishing, Paris and Luxembourg
6. Bonomi A. 2017, A Technological Model of R&D Process, and Its Implications with Scientific Research and Socio-Economic Activities, *IRCrES Working Paper*, 2/2017.
7. Arthur B. 2009, *The Nature of Technology*, Free Press, New York
8. Bonomi A. 2019, The Start-up Venture Capital Innovation System, Comparison with industrially financed R&D projects system, *IRCrES Working Paper*, 2/2019
9. Haour G. 2004, *Resolving the Innovation Paradox: Enhancing Growth in Technology Company*, Palgrave Macmillan, St. Martin Press LLC, New York
10. Chesbrough H.W. 2003, *Open Innovation: The New Imperative for Creating and Profiting from Technology*, Harvard Business School Press, Boston, MA

11. Bonomi A. 2017, A Mathematical Toy Model of the R&D Process, How This Model May be Useful in Studying Territorial Development, *IRCrES Working Paper*, 6/2017

12. Scherer F.M. Haroff D. 2000, Technology Policy for a World of Skew-Distributed Outcomes, *Research Policy*, 29 (4–5), 559–566

Perspectives and Future of Technology Innovation

9

In this chapter are discussed the perspectives and scenarios of a possible evolution of the technology innovation system and of technologies with a high potential of development. These perspectives are based on the described models of organizational structures for innovation, and in particular on the possible evolution of the industrial platform system in a form of a platform network with the presence also of secondary platforms. Furthermore, a scientific approach is presented to technology forecasting and are discussed the technology sectors that present the major possibilities of development, considering their technological combinatory potential or increased availability and exploitation of new phenomena discovered by science. Finally, are discussed some intrinsic dangers of future evolution of technology concerning the negative social impacts in particular of AI and ICT.

9.1 EVOLUTION OF THE TECHNOLOGY INNOVATION SYSTEM

The history of development of the technology innovation system has shown three main transitions on how technology innovation is organized. The first system, based on industrial R&D projects, found its origin around 1870 [1], and it was accompanied, in the first decades of the XX century, also by formation of organizations supplying R&D services to industry [2]. The second transformation occurred after the Second World War with the appearance of new important actors for R&D activities such as universities, public and private laboratories and the formation of the SVC system, in what it has been called the *distributed innovation* DI system [3]. At the beginning of the XXI, century there was the formation of a third new organizational structure for innovation, different of existing structures, and involved in the increase of knowledge and not about new strategies of financing. This new structure is characterized by a new system of relations and aggregation of firms called industrial platform [4], and derived by a new form of business based on relation among various type of actors, including firms and consumers [5]. The development of the technology innovation system has generated inclusive and not necessarily alternative systems. In fact, R&D projects are present in startups activities, and industrial platforms may have the presence of R&D projects and startup in their structures. As previously cited, the main differences are in financing strategies between the R&D projects system and the SVC system because VC finances technology innovations for their sale, and not for their exploitation as industrial capital. On the other side, the industrial platform system is not involved directly in financial strategies but in increasing knowledge through exchange of information between the platform and the users of a technology, and between peer producers and the platform. Such increase of knowledge is available for generation of innovative ideas for improvements and formation of new technologies. Considering an evolution of the innovation system based on the increase of exploitable knowledge, favorizing in this way more the generation of new technologies than the protection of technological knowledge and industrial property, it would be possible the formation of a new innovation system based on an *industrial platform network* (IPN). This system has potentially some technological advantages in respect to the actual DI system. It is then interesting to describe how this evolution of the technology innovation system may occur from the present DI system, and which are the advantages of a new IPN system in respect to a system in which confidentiality of technological knowledge and industrial property is prevalent.

9.1.1 The Distributed Innovation System

The DI system represents an evolution of the technology innovation system, occurred in the second half of the XX century, in which companies do not develop new technologies only under secrecy in their industrial R&D laboratories. In this way, a firm defines its high technological impact by developing and making use of external technologies, possibly more expensive but more efficient. That is possible by exploiting various types of relations concerning services and cooperation with other firms, as well as with other entities developing new technologies, enhancing growth in technology companies [3]. In this system, a company takes in consideration innovation strategies that

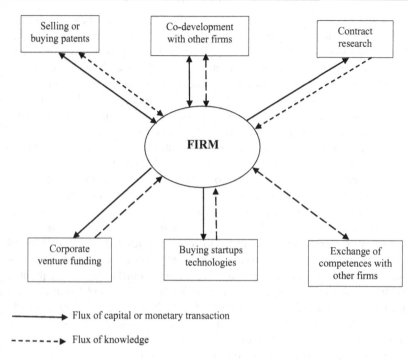

FIGURE 9.1 A schematic view of the distributed innovation system.

include not only internal R&D but also contract research with external laboratories, selling or buying licenses, co-development of technologies with other companies, acquisition or venture funding of startups, as well an open view in the exchange of competences. In Figure 9.1, we have reported a schematic view of the DI system, seen from a technological point of view, and indicating the fluxes of knowledge, capitals and monetary transactions among the various actors of the system. A characteristic of the relations of a company in the DI system is that they are *discontinuous* that is different from the case of the IPN system in which the relations are relatively *continuous* because of the activity of platforms.

9.1.2 The Industrial Platform Network

The IPN represents an evolution of the platform system applied to industrial technology innovation and resulting by a networking of these platforms supplying technologies with firms demanding technologies, in a new system of organization of fluxes of knowledge, capitals and monetary transactions.

Presently industrial platforms know a certain diffusion, but there is not available an important industrial experience allowing the development of quantitative simulation models and fully knowledge of the real advantages and limits of this organizational structure as in the case R&D and SVC systems [4]. However, technology dynamics describes some aspects of generation of innovative ideas and development of new technologies that may favor the diffusion of platform structures. In particular, it might be considered that companies, operating in certain technological sector, depend, in manufacturing their products, on availability of specific technologies that may be developed and available from other companies. These last companies might consider, for the supply and development of their technologies, to operate as a platform. On the other side manufacturing companies might consider, for their needs of technologies, to have continuous links with a certain number of platforms specialized in specific technologies necessary for the manufacture of their products. Furthermore, an industrial platform may be interested to acquire or develop, in a discontinuous way, new technologies and products with peer producers constituted by external companies, startups and research laboratories. Furthermore, peer producers may have the same type of relations with many different platforms, all that forming an industrial platform network in which supply and demand of technologies, fluxes of knowledge, capitals and monetary transactions are spread throughout the network. In Figure 9.2, we have reported a schematic view of the network considering a peer consumer firm linked to various platforms with their partners, and peer producers linked with discontinuous relations with various platforms. It is interesting to know that a primitive form of industrial platform network exists in Italian industrial districts, for example in the case of production of faucets and valves as described below.

The production of faucets, and similarly of valves, follows a technology with a structure reported in Figure 4.2 concerning the model of technology. This industry starts from raw material consisting in brass bars and ingots. Three operations of this structure: the hot stamping, the casting and the chroming are normally carried out by subcontracting industries. These subcontractors operate similarly to an industrial platform as they have multiple clients with the firms producing faucets and valves. On the other side, these firms tend to have continuous relations with their subcontractors. The great difference of this system is the practical absence of an important role of peer producers as in the case of ICT, although there is a certain increased diffusion of knowledge among firms and subcontractors due to the network. This technological structure of production in form of primitive platform network is a strength for the competitivity of the district.

Actually, it is also a possible a further evolution of the IPN system in which peer consumers might act also as secondary platforms as reported in Figure 9.3. That is possible because of the interest of a peer consumer to

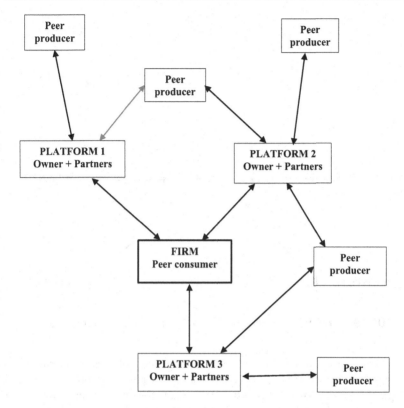

FIGURE 9.2 A schematic view of the industrial platform network.

act also as secondary platform, in respect to the final consumers, enhanced especially in the case that its products are in IoT with a useful exchange of information between the firm and the final users of the product. The platform relations with its consumer are important to obtain knowledge about the use and possible flaws of the product allowing improvements and even ideas for new technologies. A possible example of peer consumer firm acting as secondary platform in an IPN system is given as follows.

That is the case of production of electric cars with autonomous guidance. An industry producing such type of cars needs various types of technologies concerning batteries, artificial driver systems, electric motors and transmission systems. These technologies may be developed internally to the firm but might be also supplied by specialized industrial platforms and, at the same time, the car manufacturer may operate as a secondary platform with an exchange of knowledge on the functioning of the cars, as IoT products, with its customers.

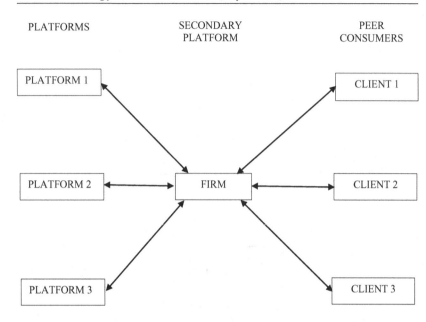

FIGURE 9.3   A schematic view of a secondary industrial platform system.

In this case, the internal development of technologies of a firm enters competition with the supply of technologies by the platforms. From the technological point of view the IPN system is more efficient because there is a higher generation and exchange of knowledge available for innovation, in respect to an integrated system of development of its own technologies. Of course, business or strategic considerations could also consider more favorable internal developments instead of use of platforms for the technological needs.

9.1.3   Advantages and Limits of an Industrial Platform Network

The formation of an IPN system entering competition with the present DI system is characterized by advantages concerning the generation and exchange of knowledge, and limits concerning the necessity of aggregation. In fact, the fundamental advantages of the IPN system is in the generation and diffusion of technical knowledge through the relations between generators and users of technologies. Such relation favorizes solution of problems during the use of a technology, reduction of time of the learning curve in reaching optimal conditions of use of a new technology, improvements in the use of a technology and generation of innovative ideas typically for incremental innovations. Such advantages derive from the combinatory nature of technology innovation and the availability of knowledge that is maximized in the IPN system. The factors limiting aggregation in the IPN system depend on the chosen technological strategies by companies. These factors depend on the degree of importance of industrial property and confidentiality, characterized by internal development of technologies, and conservation in secrecy of the knowhow. A firm operating in a DI system, in deciding to become a platform, should consider whether its position as supplier of a specific technology is sufficiently strong to establish continuous relations and exchange of knowledge with peer consumers interested to its technology. A firm operating in a DI system, in deciding to become a peer consumer, should consider whether it is of interest to obtain specific technologies from platforms with continuous relations, abandoning efforts to develop internally some specific technologies, concentrating efforts in developing the product concept, combining the various technologies supplied by the platforms, ready to cannibalize previous products if more useful technologies are available from the platforms. Concluding it shall be remarked, as previously noted, that the described transformation of the DI system into an IPN system is not a forecasting of the evolution of technology innovation activity, but rather a description of a possible scenario whose possibilities of realization depend on economic and business factors, and whether a strong technology development with its positive impacts would be preferred than industrial property exploitation in the business.

9.2 FUTURE OF TECHNOLOGIES

Following the studies on technology dynamics the driving force of technological developments is represented in particular by the combinatory potential existing for the generation of new technologies joined with the future availability of new or not still exploited phenomena discovered by science [6]. These factors may originate an exponential growth of new technologies because of the combinatory nature of the processes in presence of a continuous availability of science discoveries [4]. Looking to the present scientific and technological reality, we have made a choice about the specific technological sectors that, in our opinion, have a high potential in the future development of technologies and that are as follows:

- AI
- Synthetic biology
- Nanotechnologies
- Applications of quantum physics

The driving forces of development of these technological sectors are: the enormous amount of possible applications in the case of AI, the enormous amount of combinatory potential in the formation of DNA with natural or artificial molecules in the case of synthetic biology, the enormous amount of materials that may be reduced to a nanometric dimension with new exploitable properties in nanotechnologies, and existence of numerous quantum physics phenomena that are only at the beginning of their exploitation.

9.2.1 A Scientific Approach to Forecasting

In considering the future development of technologies previously cited, it is important to define which are the effective possibilities of forecasting of their evolution. We have already cited, discussing the relations between technology and the environment, the existence of limits of technological objectives based on scientific considerations. In fact, a scientific approach to forecasting technology evolution may be found in a past book on foreseeable future [7] written in 1957 by Georges Thompson, Nobel Prize in physics for the discovery of the undulatory nature of the electron. The interest of this book is twofold. The first concerns the fact that this forecasting has been made more than 60 years ago allowing an effective comparison between the made forecasting and the present technological reality. The second concerns the fact that the author introduces, probably for the first time in technological forecasting, a scientific approach giving rules limiting the possible forecasting by respect of fundamental scientific laws. The author cites in fact seven principles limiting the technological forecasting that are as follows:

- The constancy of the insuperable velocity of light
- Conservation of mass and energy
- Conservation of electrical charges
- Conservation of magnetic polarity
- Heisenberg principle of indeterminacy
- Pauli principle of exclusion
- The law of entropy

The author has considered for his forecasting various technological fields concerning for example energy, materials, transport and communication. In the field of energy, he believes in the development of nuclear energy based on uranium fission as a possible future abundant and cheap source of energy, and cites also the possible development of nuclear energy based on nuclear fusion. In the field of materials, he has a traditional approach not imagining the different exploitable properties of nanometric material made in nanotechnologies. In the field of communication, he bases forecasting on Shannon's theory of information with definition of bit, but sees limits in the frequencies available for telecommunications, that is not actually the case. About transport he considers only traditional technologies, nuclear energy useful only in the case of submarines, but he notes the possibility to use multistage rockets to leave the earth, but not the possibility of launching satellites that in fact in future would become very important in many fields including communication. Concerning biological applications, he considers the future of biotechnologies in particular in controlling mutations. It should be noted that at the time the book was written, the structure and function of DNA were not discovered. He recognized also the future role of mechanization and automation thinking to future applications of calculators for this purpose, anticipating digitalization of manufacturing and the decrease of routine work, accompanied by increase of planning and designing activities. Finally, he believes in the future of computers and use of the binary code and even the possibility to reproduce cybernetic artificial "animals", and realizing artificial neural systems for learning, but not the miniaturization of electronic circuits that has in fact allowed the development of modern ICT. Concluding, there is a right forecasting of some new technologies, but not for other present technologies such as synthetic biology, nanotechnology and miniaturization of electronic circuits. In fact, a nearly full forecasting of future of technology does not appear possible, that because of occurring of unpredictable casual events and discoveries that may have a strong impact on the direction of evolution of technologies, typical of the complex behavior of the technology system, and that shall be taken in account presenting the future of new technologies.

9.2.2 Artificial Intelligence

This technology involves many disciplines such as mathematics, informatics and neurosciences. Historically the evolution of AI may be divided in a first "classic" period and in the present "modern" period [8]. Classic AI corresponds to machines able to do operations for which they have been instructed, and based principally on a top-down approach in the reproduction of human intelligent behaviors. That using fundamentally the logic operation *if... then...* and further introducing fuzzy in addition to Boolean logic, attributing a value of probability to the logic assertions [8]. On this base, since the 1980s, have been developed expert systems accompanied by problem solving and data mining systems. Problem solving is an aspect of AI based on the input of a set of rules that can cover all the possibilities of a system, and obtaining an output for the searched solution. The limits of expert systems concerned the difficulty to collect all the necessary rules to operate in a complex reality. In order to improve these capacities, it has been necessary to develop a modern form of AI based on artificial neural networks, and not simple logic circuits. Differently from classic AI the approach in modern AI is of bottom up type. It includes some fundamental elements of intelligence that are assembled, carrying out a learning phase for a certain time, and after evaluating the validity of the results [8]. Practically modern AI tries to imitate the human system of learning based on a network of neurons existing in the brain. After a certain period of learning, there forms a behavior in which the network gives efficient answers to the various stimulus favoring impulses corresponding to an elevated fitness. In this way, a modern AI machine may face situations for which it would not be possible to forecast a priori all the necessary cases for a top down programming typical of classic AI. It shall be noted that, while with classic AI machines, it is possible to know exactly the used logic steps of the machine, with modern AI, it is not possible to know what they have exactly learnt, and it is necessary to verify in practice the capacity to carry out a certain task as it is done for a person. Finally, we may cite in the field of neural networks the existence of studies on connection of electronic circuits with the neuron network of the human brain, and also the opposite with the culture of biological neuron networks connected to electronic circuits. The first case is studied especially with the aim to eliminate human disabilities. The second case, that it is only in an initial phase of development, the formed network by multiplication of neurons may be stimulated and connected to electrodes. The collected outputs may be addressed to actuators forming in this way intelligent biological machines [8]. The modern form of AI has an enormous potential of applications in many fields including its use in manufacturing industry allowing the development of enabling technologies that have been described, for example, in an Organisation for Economic Co-operation and Development (OECD) report [9]. Concluding about the future of AI it is of interest to discuss also the dilemma whether the development of AI may reach or not a level similar or even superior to human intelligence. This possibility is sustained in particular by Ray Kurzweil that foresees an explosion of AI surpassing largely the human intelligence, in what he called a *technological singularity* [10]. A second position considers that AI may not be equivalent to human intelligence because of the presence in the brain of phenomena not reproducible by an artificial machine. That is the case of Roger Penrose, Nobel Prize in physics, that sustains the idea that in human intelligence is possibly present a quantum physics process that he has hypothesized in a famous book [11]. Another supporter of this second position is Federico Faggin, an Italian physicist that has invented the microprocessor working at INTEL, supporting in a book that the awareness is a specific inimitable characteristic of the human brain [12], a view that might be correlated also to the field of quantum physics with the not deterministic characteristics of nature. Actually, there does not exist any scientific demonstration that sustains the various hypotheses of these authors. Considering the model of human cognitive development elaborated by Jean Piaget [13], the full cognitive capacities are reached in humans at 15–16 years of age. It should be noted that the large learning experience cumulated by human brain, although this amount of knowledge is clearly inferior to memory capacities of AI machines, it is however highly differentiated, and that favorizes the intuitive not rational process of creativity necessary, for example, to generate innovative ideas through unconscious combinatory processes. It may be argued whether an AI machine, that needs in every case a period of instruction through the experience, might reach the same conditions of human cognitive experiences in less time of that necessary for humans. In fact, if it is true that AI machines have much higher speeds of memorization and processing of information than human brain; however, the limiting step during instruction is in many cases the occurring time of phenomena that should be memorized and processed, and not the time of memorization and processing, and that is the same either for humans or machines. That would possibly make necessary for AI machines instructions times of many years to reach the equivalence of human intelligence. A situation that may raise problems about their interest and economy. Another question is on the fact that it is not clear which would be the advantage for the humanity to try to develop a technological singularity, i.e. machines with autonomous self-construction ability, autonomous energy supply and consciousness, for which we cannot really forecast their behavior, and that might have dystopic consequences in the human social system, as discussed later about intrinsic dangers of technology evolution.

9.2.3 Synthetic Biology

Synthetic biology is in fact a branch of biotechnology. It is supported by a more and more increased knowledge of biological processes, able to develop technologies with a high degree of radicality and with a great economic and social impact [14]. Biotechnologies are generally based on the use of phenomena concerning natural processes and molecules, for example by introducing specific genes of an organism into the DNA of another organism producing a genetically modified organism (GMO). In the case of synthetic biology, the modifications of DNA are designed producing actually a genetically synthetized organism (GSO) for a specific application. The great development potential of this technology is in the enormous number of possibilities linked to the possible DNA design, and also to the possibility to substitute natural components of DNA with artificial bases that may produce artificial proteins containing amino acids different from the natural ones [15].

Typically, the method consists in the elimination in a simple organism, such as bacteria or yeasts, the part of DNA that does not concern the vitality of the organism, and after adding a designed part of DNA with the objective to implement in the organism a specific wanted function. This function is implemented by the use of a series of enzymes and biological active molecules in order to start, catalyze and terminate the process. The system realizing the expected tasks for the microorganism consists of various parts, typically DNA parts that codify the instructions, and devices consisting of elements that implement the required biological functions. The system is designed following similar methods used in electronic circuits, using specific symbols representing the various parts of functioning. The great difference is that in synthetic biology, different from electronic circuits, the parts are not physically connected but are present in the cellular medium acting thorough molecular or macromolecular processes. For this purpose, it is considered also the use of AI to forecast at computer the function of certain biological circuits. Present research effort in synthetic biology are involved in particular in CRISPR gene editing with the possibility to cut DNA at the desired location allowing removal of existing genes and adding new ones. Potential applications of synthetic biology are extremely wide, and since a certain time it exists for example the microbiological production of insulin, but there are applications realized or under study concerning pharmaceutical products, chemical products, proteinic materials and tissues, biosensors, biofuels, etc. It shall be noted that certain biomolecular processes involving also synthetic biology may be used as logic circuits. That open the possibility to realize through synthetic biology what may be considered a biological computer system [16]. This system might be connected to biological sensors and give instructions to biological or electronic actuators. Although presently such biological computing system has difficulties in its reliability and reproducibility, but there are not reasons that these problems would not be overcome as in fact these types of circuits exist and are functioning in nature. Further applications under development concern the degradation of mRNA of virus, gene editing substituting the use of antibiotics and GSO animals able to eliminate detrimental insects.

9.2.4 Nanotechnologies

With the name nanotechnology, it is intended a complex of activities concerning scientific research (nanoscience), and technological applications based on particular phenomena existing in matter reduced to a nanometric dimension. In fact, the physical and chemical properties of matter with a given chemical composition are different, in respect to those existing in a bulk dimension, when nanometric domains and particles are considered. These properties may be possibly exploited for new technologies. Nanotechnologies concern not only exploitation of properties of nanoparticles but also of other nanometric forms, such as nanotubes, nanometric layers or deposits on a surface, and nanocomposites. These nanocomponents may be used also in the design of molecular machines, existing in fact in the living matter, and that might find various applications [17]. The development of nanotechnologies has been possible also by availability of new microscopic instrumentation such as TEM, SEM, AFM and STM, as well as new assembling methods at the nanometric level. All that has allowed to investigate and manipulate directly particles and layers of nanometric dimension, and to discover new exploitable phenomena for applications. In the case of STM, it is possible also to manipulate directly atoms, such as atoms of xenon, as well as small molecules such, as CO (carbon monoxide), deposited on a suitable metallic surface at very low temperatures, that might find future-possible applications. The present sectors in which nanotechnologies have given major innovative contributes are those concerning new nanostructured materials, for either functional or structural applications, and composites containing nanoparticles imparting specific, mechanical, optical and biological properties to a broad range of different materials. Another field in which nanotechnology imitates biological systems is the development of selective and efficient catalysts, e.g. for the energy sector (photocatalysts for hydrogen production, CO_2 capture and recycling). Nanotechnology developments have also an impact on next industrial transformation in terms of product design and performance [9]. Another important application of nanotechnologies may be in the field of sensors for green technologies with sensibility even at atomic or molecular level [18]. Another one is in digitalized productions, for example in the technology of digital twin consisting in a digital representation of a physical process based on feedback of sensors of the real system [19]. Another important field of application of nanotechnologies is in electronics and then in digitalizing technologies. In fact, nanotechnologies have become important in particularly with the development of miniaturization of circuit now arriving close to a molecular dimension. Finally, nanotechnologies are involved also in the sector of biotechnologies for medical applications through diagnostic and pharmaceutic products using special nanometric formulations to enhance biological activity and the targeting of drugs, and imaging media to specific biological targets.

9.2.5 Quantum Physics Applications

From the technological point of view, quantum physics is important because it is an interesting source of many phenomena, very different from those based on classic physics currently exploited for technology innovations.

The relatively recent developments of quantum physics may indicate that is still far the exploitation of all its potential applications. It is not our intention and possibility to explain all the known quantum phenomena exploitable from a technological point of view, but just to give an indication, presenting also some limits and difficulties existing in their technological exploitation. Before presenting some quantum phenomena interesting technological applications, it is useful to give, in a very simplified manner, some important aspects of quantum physics that differ from classic physics. The first important postulate existing in quantum physics is represented by the *Heisenberg principle*. This principle limits the accuracy that it is possible to have in the microscopic world establishing that it is not possible to determine at the same time the exact velocity and the position of a particle, and also the exact measurement of energy and time of a particle, putting limits to what we may really know about a microscopic system. Another important mathematical description in quantum physics is the *wave function* that is able to establish the state of a quantum system. The values given by the wave function may be related to the probability to obtain a certain result of measurements made on the system. The description of the wave function and how it may evolve with time is given by the *Schrödinger equation*. In the case of the hydrogen atom, constituted by a proton as nucleus and an external electron, the Schrödinger equation may calculate the possible various energy levels of the electron, and values of the wave function in relation with the probability of position of the electron. It shall be noted that the values of the wave function are not actually related to the probability of position of the electron, but on the probability of the resulting position when it is measured. That opens the problem about the nature of measure in quantum physics, and the fact that we cannot have a physical image of a quantum system as we may have one for a macroscopic system. A final important aspect of quantum physics is the existence of entities that may have two completely different natures. For example, an electron may have a behavior as a particle with its specific mass and electric charge, but also as an electromagnetic wave with its own frequency presenting phenomena of interference. In the same manner, electromagnetic waves have not only the nature of wave but also that of packets of energy (quantum) called photons. In fact, in quantum physics, are possible experiments determining only one of the natures of the entity, and it is not possible to imagine an experiment that verifies both natures of the entity at the same time. In quantum physics, there are also properties that do not have really an equivalent for macroscopic bodies. That is the case of *spin*, that is actually related to magnetic properties of matter, and metaphorically described as representing one of the two directions of rotation of a particle. This property is for example used in medical instrumentation based on magnetic resonance. A fundamental phenomenon existing in quantum physics is the *quantum superposition*. It is based on the fact that two or more quantum states may be added together, and then superposed, forming another valid quantum state. Such property is exploited for example in quantum computers that are realized using superposed states of a quantum property corresponding to a quantum logic state called *qbit*. In fact, that is equivalent to superimposition of both 1 and 0 states in computerization actually not allowed by classic logic. Physically this situation may be realized, for example, with an atomic nucleus on which are superposed two spin states of values ½ and -½. These superposed logic states may be exploited to increase enormously the computing potential through suitable algorithms with applications for example in cryptography. Other applications of quantum computer under development concern the simulation of complex physical and chemical systems, development of materials and in customized patient medicine. An important quantum physics phenomenon is that of *decoherence*. It is represented by the loss of coherence of a quantum system because it cannot be completely isolated and then exposed to external influences on its state. Measurements in quantum systems are a typical cause of decoherence, and a measurement of a superposed quantum state will result in the appearance of one of the superposed states following the probability related to the wave function of the system. Quantum decoherence, more than an exploitable phenomenon, represents sometime a limit in the exploitation of quantum applications. For example, in the case of quantum computers, decoherence limits the time in which the computer is coherent and available to make the calculations. In the case of the present types of quantum computer prototypes, the time available for calculation in which it is conserved, the coherence of the computing system is around 100 microseconds, and that is possible by conserving the system at very low temperatures with liquid helium to limit interactions causing the decoherence. However, for quantum computing are under development also systems based on photons with the role of qbits. These systems, using optical fibers, do not necessitate to work at low temperature with the exclusion of some final components of the photonic computer. Another important quantum phenomenon is the *entanglement* that establishes a special correlation between two particles or photons, emitted simultaneously in a quantum system, and having opposite values of their quantum property. There was a controversy between Einstein, believing that the opposite properties values were assumed at the moment of the emission, and Bohr, believing that the properties were in superimposition assuming a value only at the moment of its measure, forcing consequently the other particle or photon to assume the opposite value whenever it was measured. Experiments carried out later, based on a suitable statistics for this phenomenon, showed correct the Bohr's position. That was shown also making the second measure very far in order to show that a hypothetical message to the other measure, in order to assume the opposite value, should be transmitted at a speed higher than the velocity of light. Such phenomenon of entanglement is studied for potential applications for example in cryptography and even for teleportation. Finally, there are three important phenomena with current technological applications that are the *tunneling effect*, the *photoelectric effect* and *superconductivity*. The first one represents the possibility of a particle, for example an electron, to cross (tunneling) an energetic barrier even if its energy is lower than the barrier, that by effects of its undulatory properties. This effect is used in electronic circuits applications. The second one is based on the interaction of photons with matter. That is possible only if the energy of the photon is enough high to make the interaction. The inversion of this effect may be used in extracting electrons from a metal and directing them on a target producing X rays for investigations of the structure of matter. Superconductivity, a quantum phenomenon of electrons obtained at very low temperature in special alloys, makes possible the obtention of very high magnetic fields, necessary for example in medical instrumentations such as MRI and PET and in plasma nuclear fusion. Concluding, the potential of application of quantum phenomena appears very numerous although there are limits in certain case because of the decoherence effect. That makes necessary, for the exploitation of certain quantum phenomena, to operate at very low temperature and then in presence of equipment supplying liquid helium for cooling. That makes impossible a miniaturization of the device as it is possible for electronic circuits. However, it shall be noted that this limitation is not present in the case of use of quantum phenomena linked to photons. That is the case, for example, of the laser and partly in the case of photonic computer.

9.3 INTRINSIC DANGERS OF TECHNOLOGY EVOLUTION

In the previous chapter concerning the nature and concept of technology innovation, we have already discussed of dangers associated with the evolution of technology. These dangers have been raised in particular by a German philosopher, Martin Heidegger, and by his scholars Hans Jonas and Hans Georg Gadamer discussed previously. Direct dangers of technology, concerning in particular the environment, have been raised by Hans Jonas, and discussed in detail in the previous chapter about relations between technology and the environment. Martin Heidegger cited intrinsic dangers of technology in terms of pushing humans to develop technologies in which technology is the main objective, and not its use. Hans Georg Gadamer, on the contrary, found as main intrinsic dangers of technology in the modern technologies of communication with possible manipulation of public opinion. In a discussion of intrinsic dangers of technology evolution, it is necessary to consider in particular the negative social impact of technologies, because of the availability of enormous amounts of power and knowledge, and of the modern communication technologies.

9.3.1 The Negative Social Impacts of Technology

An interesting view about the social impact of technology may be found, not only in the reflexions of philosophers previously cited, but also in those of scientists, in particular in the case of Werner Heisenberg, a well-known German physicist, Nobel Prize for his contribution in the foundation of quantum physics. He considered that the position of humans in front of the nature has been now radically changed at the light of the modern physics, including technology being it the result of the scientific advances, and reported his views in a book published in 1955 [20]. This scientist considered the fact that the big changes occurred in our environment and way of life, induced by the rapid and radical evolution of modern science and technology, have also transformed dangerously our way of thinking, without leaving time to the humanity to adapt to the new conditions of life. He cited for this purpose critics to technology made by Chuang Tse, successor of Lao Tse the mythic founder of Taoism in the III century AC. Chuang Tse exposed its contrariety to technology in a short story, reported by Heisenberg in his book, that may be summarized as follows:

Tse Cung reached a territory and looked at an old farmer that moved manually and with difficulty a bucket of water from a well to an irrigation channel and asked him why he did not know the use of a lever with a bucket moving water with much less strain. The old farmer looked at Tse Cung and told

> my guide has told me: if one use machines he does all his acts mechanically, when one does all his acts mechanically finally he has a mechanical hearth, loses the simplicity and becomes insure in the motion of his spirit, of course I know this machine but I will be ashamed to use it.

Of course, we may argue sustaining the utility of machines for the wellness of humanity, but there is a lesson in this brief story that may be hold in the era in which technology does not increase only the physical possibilities of humans but also their intellectual capacities. In fact, the danger of use of technologies increasing the intellectual capacities may be found in a tendency of humans to transfer fully to machines with AI the answer to questions and solution of problems, as well as management of activities, with the danger to reduce human knowledge to simple procedures consisting in a sequence of touches on a keyboard or a screen, the mechanical acts cited by Chuang Tse.

The observations made by Heisenberg are similar to those made by Heidegger about dangers of technology discussed previously. In fact Heisenberg was cited by this philosopher in his essay on the question on technology [21]. Actually, it should be noted that both Heisenberg and Heidegger reached these conclusions considering only the impact of availability of enormous power such as that from atomic energy. That before the development of ICT making also available an enormous amount of knowledge, not necessarily true and useful, worsening the problem of human adaptation to these new conditions. Actually, the today's question on technology concerns the submission of humans to decisions made by machines with their intrinsic uncertainty, accompanied by a lacking of knowledge about how machines work in many people that use them. In fact, we may pose questions whether the machines with AI are really suitable and reliable to accomplish fully the complex tasks for which they could be developed. If a machine, such a computer, is of Turing type, based on deterministic rules, it is possible to know in principle how it works and which are the data it knows. That it is not the case of an AI machine based on universal approximation rules simulating the biological neural network, and necessitating a period of instruction to accomplish tasks of superior complexity. In this case, it is not possible to know exactly what the machine has learned necessitating a verification similar to that is done for humans. All these dangers represent also a challenge for future education of humans about what they shall and can know on machines with AI in order to control them, and avoid ignorance about how these machines operate with the possible dangerous consequences.

9.3.2 The Negative Impact of Communication Technologies

A further intrinsic danger of technology is linked to new communication systems through internet and social networks with their large diffusion and freedom. That has been anticipated in a certain way by the philosopher Hans Georg Gadamer cited previously. These technologies have been developed in the 1970s, creating data banks connected with remote computers with the aim to have a better diffusion of knowledge, in particular of scientific or technical nature. However, their development, with a large diffusion and freedom in their use, has also transformed them as a means of diffusion of ignorance in form of anti-scientific information, fake news, etc. That accompanied also by a deleterious use of these technologies in getting data unconsciously from the users, and resulting in an unaware influence with dangerous or unwanted effects on the users of this technology. Furthermore, another of the most dangerous aspects of modern communication technologies is the transfer of rational thinking, for example through the use of "likes" typical of social networks, to emotional thinking leaving rational thinking to machines with the cited dangerous consequences. Actually, it should be considered that this type of technologies has specific new characteristics in influencing the socio-economic system, not necessarily positively, beyond their technological nature. That has been noted for example by a German journalist [22], discussing differences, origins and future of the Silicon Valley views. In fact, Silicon Valley technologies, differently of common focus of industrial and financial activity, have as primary focus not on money, not on how much we consume but what we consume and how we live, and that not stumbling haphazardly into the future but rather with a clear agenda. These technologies cannot be considered simply based on internet or social networks, nor based on intelligence and data supply and messaging services, or jobs replaced by software and the launching of entire new industries. In fact, what they are producing is a societal transformation in which the digital revolution is not just altering specific sectors of the economy, but changing the way we think and live not necessarily in a positive way.

REFERENCES

1. Basalla G. 1988, *The Evolution of Technology*, Cambridge University Press, Cambridge
2. Boehm G. Groner A. 1972, *Science in the Service of Mankind, The Battelle Story*, Lexington Books, Heath and Company, Lexington, KY
3. Haour G. 2004, *Resolving the Innovation Paradox: Enhancing Growth in Technology Company*, Palgrave Macmillan, St. Martin Press LLC, New York
4. Bonomi A. 2020, *Technology Dynamics: The Generation of Innovative Ideas and Their Transformation into New Technologies*, CRC Press, Taylor & Francis Editorial Group, London
5. Cicero S. 2017, *From Business Modeling to Platform Design*, https://platformdesigntoolkit.com/platform-design-whitepaper/
6. Arthur B. 2009, *The Nature of Technology*, Free Press, New York
7. Thompson G. 1957, *The Foreseeable Future,* Cambridge University Press, Cambridge
8. Warwick K. 2012, *Artificial Intelligence, The Basics*, Routledge, Taylor & Francis Group, London
9. OECD. 2017, *The Next Production Revolution: Implications for Governments and Business*, OECD Publishing, Paris

10. Kurzweil R. 1990, *The Age of Intelligent Machines*, MIT Press, Cambridge, MA
11. Penrose R. 1989, *The Emperor's New Mind*, Oxford University Press, Oxford
12. Faggin F. 2021, *Silicon: From the Invention of the Microprocessor to the New Science of Consciousness*, Waterside Productions, Cardiff by the Sea, CA
13. Piaget J. 2001, *The Psychology of Intelligence*, Routledge (Classic), Taylor & Francis Group, London
14. Freemont P.S. (editor) et al. 2016, *Synthetic Biology: A Primer*, Imperial College Press, World Scientific Publishing Co., London
15. Callaway E. 2017, Cells Use "Alien" DNA to Produce Protein, *Nature*, 551, 550–551
16. Benenson Y. 2014, Biomolecular Computing Systems: Principles, Progress and Potential, *Nature Reviews, Genetics,* 13, 422–468
17. Davis A. 1999, Synthetic Molecular Motors, *Nature,* 401, Sept. 9, www.nature.com
18. Steffi F. 2017, "Tapping Nanotechnology's Potential to Shape the Next Production Revolution" in *The Next Production Revolution: Implications for Governments and Business*, OECD Publishing, Paris, 157–239
19. McDowel D. 2017, "Revolutionizing Product Design and Performance with Materials Innovation" in *The Next Production Revolution: Implications for Governments and Business*, OECD Publishing, Paris, 215–359.
20. Heisenberg W. 1955, *Das Naturbild der heutigen Physik*, Rohwolt Deutsche Enziklopädie, Hamburg
21. Heidegger M. 2013, *The Question Concerning Technology* (English translation by W. Lovitt), Harper Perennial Modern Thought, New York
22. Schultz T. 2015, Tomorrowland: How Silicon Valley Shapes Our Future, *Spiegel Online International*, 03/04/2015

Conclusion

10

The originality of the description of technology innovation in this book is based on the use of a scientific approach to the definition of technology as a set of physical, chemical and biological phenomena that produce an effect exploitable for human purposes. This approach is independent of industrial, economic or social factors, and allows for the description of hidden or poorly considered structures and processes characterizing technology and its innovation such as the knowhow and the transfer of technology. The science of complexity offers a simplification of the study of the enormous number of physical phenomena characterizing a technology by considering these in terms of a set of technological operations. These operations are related and control the physical phenomena of technology through the parameters with their values or choices. Furthermore, technology may be described as a time-oriented structure of its operations, and technology innovation may be seen as a change of such structure. All that allows a mathematical modeling of a technology that represent in fact a description of its functioning, and of related technological processes. Technology innovation may be considered and modeled taking account that the change of the structure of a technology occurs in structures organizing fluxes of knowledge and capitals for the generation of new technologies. These structures are actually the industrial R&D projects system, the SVC (startup venture capital) system and the industrial platform system. This view of technology and its innovations is in fact built up through the use of general concepts, processes and models studied by the science of the complexity, in fact involved in the whole process of technology innovation, from the generation of innovative ideas to the use of technology as source of improvements and new technologies. The models of technology and its innovation allow also for the description of the relations between technology and the environment in view of finding solution to pollution, depletion of resources and global warming. Applications of these models concern the optimization of functioning of a technology, and the description of the technology innovation system of a territory. Furthermore, the models of technology and its innovation allow a view of the perspectives and future of technology innovation. These

DOI: 10.1201/9781003335184-10

perspectives concern the description of a possible evolution of the present distributed innovation system toward the formation of a network of industrial platforms, boosting the generation of knowledge, useful for further technology innovations. It is also possible to describe the future of development of certain technologies, such as AI, synthetic biology, nanotechnologies and quantum physics applications, based on combinatory possibilities or availability of physical phenomena not still fully exploited. Finally, some intrinsic dangers of technology evolution concerning in particular the availability are discussed, made by technology, of an enormous power coming from atomic energy, and of an enormous amount of knowledge, not all necessarily valid and useful. That might transform dangerously the way of thinking, without leaving time to the humanity to adapt to the new conditions of life, and to have the possibility of a right use of the enormous amount of power and knowledge available. Concluding, this book, through the detailed description of models, dynamics and processes of technology innovation, has also the aim to show that technology may be a solution and not necessarily a problem, and that a sustainable technology development is possible.

Appendix

Mathematical Simulation Model of the Technology Innovation System of a Territory

This mathematical model has been developed on the base of the schematic fluxes of knowledge and capitals reported in Figure 8.2. This schematic view considers the activity of the technology innovation system of a territory consisting of a certain number of R&D projects and startups, and the presence of industrial platforms. For the definition of equations, variables and parameters of the model, we may separate those concerning the flux of capitals from those concerning the flux of knowledge. The use of new technologies needs industrial capitals and will produce ROI. The resulting industrial investments in R&D projects are however not linked to ROI but depend on other factors linked to industrial strategies. On the other side, new technologies, generated by startup with investments by VC, produce, through the exit, an ROI that it is partly retained by VC and the rest reinvested in new startups. This reinvestment contributes in this way to the total investments available for R&D and startup proposals. The generation rate of new technologies is different following R&D projects or startups. That is because of the higher degree of radicality of startup technologies, and consequently a higher abandoning rate occurring in this case. On the other side, to have a positive financial cycle for the VC activity, it is necessary to have a high ROI for technologies developed by startups to compensate for the higher rate of abandoning, and then a higher rate of formation of successful technologies generated by startups in respect to R&D projects. Looking now to the mathematical aspects, in the case of flux of capitals we have to consider that the total number N of R&D projects and startups in activity is the sum of:

$$N = N_p + N_s \tag{A.1}$$

in which N_p represents the number of R&D projects and N_s the number of startups. Introducing a factor b that determines the part of startups in the total number N defined previously, we have:

$$N_s = bN \tag{A.2}$$

and

$$N_p = N(1 - b) \tag{A.3}$$

Considering now the number of generated new technologies T, we will have a rate v for the N_p R&D projects, and a rate w for the N_s startups in which $w < v$, and then T will be:

$$T = vN_p + wN_s \tag{A.4}$$

The successful number of new technologies is S. It will be the result of a selection rate r on the number new technologies generated by R&D, and a selection rate z for technologies generated by startups, and in which $z > r$ and then:

$$S = rvN_p + zwN_s \tag{A.5}$$

Considering now the flux of knowledge, we have assumed that knowledge may be measured in terms of the number of information packages. Then R&D projects and startups generate a certain number of packages either in the case of successful development or abandoning of a technology. The number of information packages in the case of startups is also higher to take account of the generation of knowledge for business model developments. For this purpose, we consider that each R&D project generates an average number p of information packages, and each startup an average number q of information packages in which we have $q > p$. The total available number of information packages resulting from a cycle shall be increased by the available number of information packages I_i of previous cycles reduced by the fading effect represented by coefficient f. The total number of packages shall then be increased by the contribution of knowledge coming from industrial platform activities, expressed by coefficient k, and information packages coming from scientific, technical and other knowledge of external origin, expressed by coefficient e. The total number of information packages I_T available for generation of innovative ideas for R&D projects or startups proposals may be calculated mathematically by the formula:

$$I_T = \left[(pN_{p(L)} + qN_{s(L)}) + \sum_{i=1}^{n} I_i(1-f) \right](1+k+e) \qquad (A.6)$$

in which we have:

I_T : total number of information packages available for new innovative ideas after the last cycle

$N_{p(L)}$: number of R&D projects in the last considered cycle

$N_{s(L)}$: number of startups in the last considered cycle

p : average number of information packages of each R&D project

q : average number of information packages of each startup

n : number of past cycles

I_i : number of available information packages of the past cycles from i = 1 to i = n

f : rate of fading effect (*)

k : fraction of added information packages by industrial platforms

e : fraction of added information packages by external knowledge

(*) It shall be noted that for remaining information packages of past cycles, we intend that initial information packages of a cycle are reduced by fading effect f and at each successive cycle before the last one. With $f = 0$ the fading effect is not present and with $f = 1$ there is a complete loss of past information packages.

Considering that a potential innovative idea is obtained by combination of m available information packages, the generation of potential innovative ideas G is obtained by a combinatory calculation based on the number of available information packages I_T through the following equation:

$$G = I_T(I_T - 1)/m \qquad (A.7)$$

in which we have:

G: number of potential ideas for innovations

I_T: total number of information packages available for potential innovative ideas after the last cycle

m: combinatory number of information packages necessary to generate a potential innovative idea

In fact, the number G of potential ideas is a simple combinatory result, not considering any validity about specific combinations, and it consists necessarily also in a large number of invalid or even absurd combinations. It is the task of the innovation system of the territory (ISE) to find the valid exploitable innovative ideas, and the number P of effective new ideas becoming R&D projects or startup proposals. That may be obtained by introducing a rate factor s applied to the number G of the potential new innovative ideas. This rate represents a measure of the ISE of a territory obtaining in this way the relation:

$$P = sG \qquad (A.8)$$

The connection of the flux of knowledge with the flux of capitals is finally represented by the selection of R&D projects and startup proposals to be financed on the base of available R&D investments by industry, and startup investments by VC. In principle, it might be possible that the selection rate of VC might be lower with much more rejected startup proposals than R&D proposals, in order to accept proposals for technologies only with a high ROI potential; however, the model cannot calculate the total available investments in innovations from the flux of capitals because that depend also on the various strategies about financing R&D in the frame of existing different situations that cannot be taken in consideration by the model. Approximately, we might consider that the selection rate of R&D projects and startup proposals is roughly the same, and we may define a rate t determining the total number N of financed R&D and startup proposals constituting the innovation activity of the cycle following the relation:

$$N = tP \qquad (A.9)$$

It shall be noted that we consider that the rate s and the rate t are the same for both R&D projects and startups constituting in fact a simplification assumed by the model. In conclusion, we may express the total number N of R&D projects and startups carried out in a cycle as a function of generated packages of information I_T by the previous cycle combining equations (A.6), (A.7) and (A.8):

$$N = ts I_T (I_T - 1)/m \qquad (A.10)$$

It may be considered finally that the mathematical model is built on the base of the assumption that both R&D projects and startups are present in the innovation activity, that they have different conditions of development and generation of successful technologies, and that it is necessary for the flux of capitals and of knowledge to consider a specific number of N_p R&D projects and N_s startups in the innovation activity of the territory, determined by the factor b

defined previously, and included in equations (A.1), (A.2) and (A.3). The factor b represents in fact the degree of diffusion of the startup activities in a territory, and it represents an important factor determining its technological situation. Concluding the description of the mathematical model, we have reported the list of variables and the list of parameters used by the model in Table A.1.

Considering the model with its simplifications, we have chosen to study three types of scenarios. Two of them concern, respectively, the influence of the diffusion of startup activities determined by the parameter b of the model,

TABLE A.1 Lists of variables and parameters

LIST OF VARIABLES
N: total number of R&D projects and startups of a cycle
N_p: number of R&D projects
N_s: number of startups
$N_{p(L)}$: number of R&D projects in the last considered cycle
$N_{s(L)}$: number of startups in the last considered cycle
T: number of formed new technologies
S: number of formed successful technologies
I_T: total number of available information packages after the last cycle
G: number of potential innovative ideas
P: number of innovative ideas for R&D projects and startup proposals
List of parameters
b: rate of startups on the total number N
v: rate of new technologies formed from R&D projects
w: rate of new technologies formed from startups
r: rate of formation of successful technology on number of technologies formed by R&D projects
z: rate of formation of successful technology on number of technologies formed by startups
p: average number of information packages of each R&D project
q: average number of information packages of each startup
n: number of past cycles
I_i: number of remaining information packages of past cycles from $i = 1$ to $i = n$
f: rate of fading effect
k: fraction of added information packages by industrial platforms
e: fraction of added information packages by external knowledge
s: rate of the innovative system efficiency
t: rate of selection of R&D projects and startup proposals
m: combinatory number for generation of potential innovations

and the influence of the diffusion of industrial platforms seen as an increase of available knowledge determined by parameter k. These two scenarios are compared with a third scenario of reference in which the innovation activity is limited to only R&D projects, and determining in these three cases the number of formed new successful technologies.

A.1　VARIABLES AND PARAMETER VALUES

In order to make calculations, it is necessary to establish the range used for the variables and the value of the various parameters used in running the model justifying the choices. The core of the model is represented by the number N represented by the sum of the number N_p of R&D projects and the number N_s of the startups active in a cycle following equation (A.1). In order to study the effects of the innovation activity including the SVC system or the industrial platform system, it is considered an initial number N for the first initial cycle by calculating the formed successful technologies, followed by calculations for the following cycles. As R&D projects and startups have different rates of abandonment, and their obtained technologies have a different rate in becoming successful technologies, it is necessary to take in consideration separately the number of R&D projects N_p and the number of startups N_s in the activity using coefficient b and equations (A.2) and (A.3). In our example of calculation, we have considered a coefficient b between 0 and 0.2 (20%). Concerning the rate of abandonment, the rate of success of R&D projects and startups becoming a new technology, they are different, as previously indicated, because of the higher degree of radicality of technologies developed by startups. Taking account of these considerations and experience in R&D, we might estimate an indicative value for parameter v of 0.0025 (0.25%), and an indicative estimation for parameter w for startups reaching the exit of 0.015 (1.5%). The values of these parameters should be considered however the most uncertain of the model. The total number T of technologies formed in a cycle may be then calculated using equation (A.4). Considering now the rate of formation of successful technologies, we might have indications in the case of R&D projects from studies about distribution values of patents, and it is possible to estimate a value of r of 0.2 (20%) for obtaining successful technologies by R&D projects. Concerning the success of technologies coming from startups, we may consider that most of the new obtained technologies are successful, resulting after a high rate of abandonment, and all characterized by a high degree of radicality, we have then assumed $z = 0.9$ (90%). The number of total successful technologies obtained in a cycle may then be calculated using equation (A.5). Considering now parameters concerning the flux of knowledge in the model, we have already established that this knowledge may be measured indicatively by the number of information packages circulating in the flux. Quantitative data on generation of information packages by R&D projects and startups, as well as the number of packages necessary for the combinatory calculation of generated innovative ideas, may be estimated approximately by experience in R&D projects and startups. They cannot be for a single project or startup a very high number. For this reason, we have considered, in a conservative view, a number of 3 information packages for parameter p associated with each R&D project, and a value of 4 for q in the case of startups in order to take account also of knowledge in business model developments. For parameter m, following the same experience, we have considered an average number of 2 for information packages necessary for the generation of innovative ideas. Another parameter necessary for calculation concerns the fading effect on information packages generated in past cycles and including past industrial platforms and external information. It has been considered indicatively that about 50% of past information is lost at each cycle. That means the total number of past information packages is halved at each cycle. The fading effect is then estimated to a value of 0.5 (50%) for parameter f. The amount of knowledge made available from the industrial platform activity depends on the diffusion of this organizational structure in the innovation system of a territory, and we have estimated that the available information may vary from a fraction k of 0 (0%) to 0.2 (20%) to be added to the available information packages. The external contribution of information packages coming from scientific, technical or other types of knowledge to the total information packages available for generation of R&D projects cannot be very high in respect to knowledge generated by R&D projects and startups. For this reason, we suggest for the external contribution an indicative added fraction e of 0.1 (10%) of the total information packages available. The total number of information packages available after a previous cycle I_T may then be calculated using equation (A.6). The number of potential innovative ideas G may be calculated using the combinatory factor m following equation (A.7). The number P of R&D projects and startup proposals to be submitted for financing depends on the ISE of the territory in exploiting the total available knowledge, and is expressed by coefficient s used in equation (A.8). In our calculation, it has been established s equal to 0.0025 (0.25%). The total number N of R&D projects and startups entering in activity will depend on parameter t for the selection of proposals and is determined using equation (A.10). For a further simplification of our calculations, we have considered, in running the model, a drastic approximation assuming $t = 1$ (100%), and then that there are always available capitals financing all valid R&D and startup proposals. Finally, it is necessary to define two variables for the parametric

TABLE A.2 Values of parameters used for the application of the model

b: rate of startups on the total number N	0–0.2
v: rate of new technologies formed from R&D projects	0.025
w: rate of new technologies formed from startups	0.015
r: rate of formation of successful technology from R&D projects	0.2
z: rate of formation of successful technology from startups	0.9
p: average number of information packages of each R&D project	3
q: average number of information packages of each startup	4
f: rate of fading effect	0.5
k: fraction of added information packages by industrial platforms	0–0.2
e: fraction of added information packages by external knowledge	0.1
s: rate of the innovative system efficiency	0.0025
t: rate of selection of R&D projects and startup proposals	1
m: combinatory number for generation of potential innovations	2

study of the model: the initial number N_o of R&D projects and startups for the first cycle, taken equal to 100, and the number of cycles considered to calculate the number of obtained successful technologies, and established in five cycles. In order to study the effects of values for parameters b and k, we have considered three scenarios:

1. Only R&D projects activity $b = 0$ and $k = 0$
2. Only R&D projects and startup activity $b = 0.2$ and $k = 0$
3. Only R&D projects and industrial platforms activity $b = 0$ and $k = 0.2$

In Table A.2, we have summarized, respectively, the values and ranges of the various parameters used for application of the model.

A.2 RESULTS OF CALCULATION WITH THE MODEL

The calculations have been made using an Excel® sheet that is reported with an example of calculation in Figure A.1. We have carried out three runs in order to see the difference of effect of presence of startups or of industrial platforms in the innovation activity in a territory, compared with the case of existence of only R&D projects activity. As cited previously, a first run of reference has

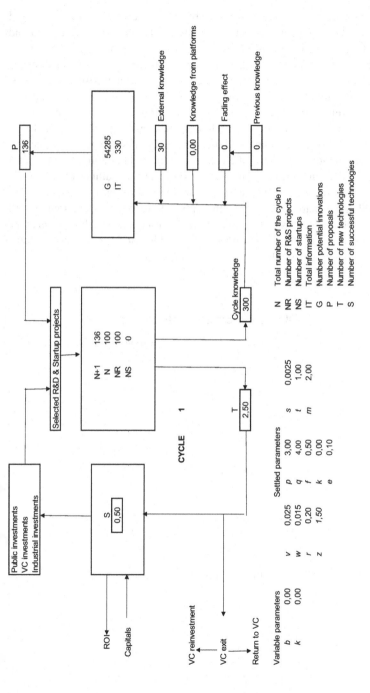

FIGURE A.1   Mathematical simulation model represented in an Excel sheet.

been carried out assuming for the variable parameters the values $b = 0$ and $k = 0$, the other two runs, respectively, with $b = 0.2$ and $k = 0$, and $b = 0$ and $k = 0.2$. The calculation has been carried out, as cited previously, for five cycles noting the formation of the number of successful technologies at each cycle. The mathematical model calculates the number of successful technologies with decimals that we do not have approximated to units in order to appreciate the differences in the various cases. Such representation of the number of successful technologies with decimals may be considered to be linked to a probability of formation of new and then successful technologies estimated by the simulation model. Actually, the values are reported in Figure A.2 only until the fourth cycle; in fact, in the fifth cycle, the obtained number of successful technologies is very high and out of scale in all the three cases. That is due, as discussed in the chapter concerning this application, to the exponential growth of the number of successful technologies as a consequence of the combinatory process of the increasing availability of knowledge, not limited by the availability of financing capitals.

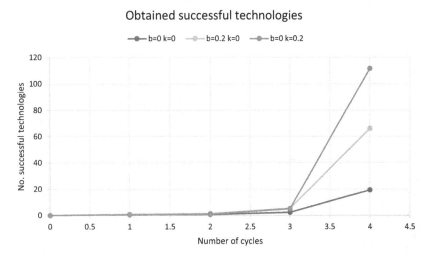

FIGURE A.2   Number of obtained successful technologies after four cycles.

Index

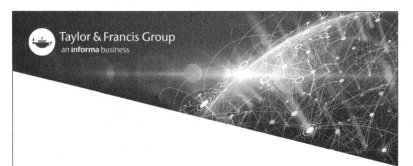

Printed in the United States
by Baker & Taylor Publisher Services